University of Michigan Museum of Anthropology
Technical Report 23

A Vertebrate Faunal Analysis Coding System

with North American Taxonomy and dBase Support Programs and Procedures (Version 3.3)

by

Brian S. Shaffer
and
Barry W. Baker

Ann Arbor
1992

© 1992 The Regents of The University of Michigan
The Museum of Anthropology
All rights reserved

Printed in the
United States of America

ISBN 0-915703-28-9

The paper used in this publication meets the requirements of the
ANSI Standard Z39.48-1984 (Permanence of Paper)

TABLE OF CONTENTS

LIST OF TABLES	v
FOREWORD	vii
PREFACE	ix
ACKNOWLEDGMENTS	xi
INTRODUCTION	1
CODING FORM CRITERIA	1
PROPOSAL FOR A NEW DATABASE CODING SYSTEM	2
Element Lists	5
Taxonomy	5
THE CODING FORM	6
COMPUTER APPLICATIONS	8
USE OF THE CODING FORM	9
Provenience Information	9
Taxon	9
Element	10
Common Elements	10
Miscellaneous Elements	11
Portion of Element	11
Mandible	11
Long Bone	12
Podials and Miscellaneous Portions of Elements	12
Side	12
Age Criteria	12
Age	12
Taphonomy	13
Weathering	13
Burning	13
Cut Marks	13
Medical Disorders/Trauma	13
Comments	13
SUMMARY	14
APPENDIX I: VERTEBRATE FAUNAL ANALYSIS SYSTEM CODES	15
APPENDIX II: dBASE FACS SUPPORT PROGRAMS AND PROCEDURES	77
INTRODUCTION	79
CHECK PROGRAM	79
XTOT PROGRAM	89

NUMERIC-TO-TEXT REPORT GENERATION
   (LINKFILE Program) 93
CONDENSE FILE PROCEDURES 99
DATABASE LINK PROCEDURES 101
REFERENCES CITED 103

LIST OF TABLES

Table 1 (Sample Data Set) 7

Table 2 (CHECK Program) 80

Table 3 (Field Name Abbreviations) 86

Table 4 (Error Key for CHECK Program) 88

Table 5 (XTOT Program) 90

Table 6 (LINKFILE Program) 95

Table 7 (Report Form Field Information) 96

Table 8 (Converted Data from Table 1) 97

Table 9 (dBase File Condensing Steps) 100

Table 10 (Data File Linking Procedures) 102

FOREWORD

Choosing a computer coding system is so important to the ultimate success of a faunal analysis that several questions should immediately be raised before adopting a coding system. Will the effort spent in encoding the data with this system really save me effort later in the analysis? Is the system readily learned? Will the coding system allow me to go where my intellect dictates without locking me into a specific way of analyzing the data? Is the coding system flexible enough to allow me to modify it to encode specific information required in my analysis? Considering the Faunal Analysis Coding System (FACS) introduced here, the answer to all four questions is an unqualified "yes."

The coding system presented here was developed by Brian S. Shaffer and Barry W. Baker (Shaffer and Baker 1990) when the research staff at the Zooarchaeology Laboratory, Texas A&M University undertook the analysis of the Alabonson Road faunal assemblage consisting of more than 128,000 specimens (Baker et al. 1991). Over a two year period, several more analyses were completed. These projects involved fauna from several localities in the southern and southwestern United States. These assemblages consisted of remains ranging in states of preservation from poor to excellent, with samples varying from a few hundred to several thousand specimens. Those who have used FACS during this developmental period have found that it is a very functional and comprehensive coding system.

The features of FACS that make it a highly adaptable system are presented in the accompanying text, but one feature I would like to emphasize is its flexibility. The analyst can choose to record the information in as simple or as detailed terms as the analyst prefers. Thus, the researcher can select only those coding options from the more than two thousand available in FACS that are pertinent to the project at hand. With repetitive use of these options the analyst soon becomes familiar with commonly used codes.

The functional features of FACS that appeal most to me are the use of standardized data manipulation software such as dBase, and the features that permit the generation of tables with proper names almost immediately after the entry of data. Having the flexibility of a data base system at one's disposal greatly facilitates the manipulation of the data in a wide variety of ways, and this in turn allows the analyst to "see" the data in as many ways as possible. Finally, the feature of having the computer generate the final drafts of tables with all labels in place and properly spelled, and all columns and rows properly tabulated is a luxury we all can appreciate.

D. Gentry Steele
Department of Anthropology
Texas A&M University
College Station, Texas

PREFACE

The use of computers in the recording of complex data, such as archaeological data, has greatly facilitated the analysis process in virtually all sciences. The computer system presented here has been created to aid in the analysis of faunal remains. We have included not only a description of the programs and support files, but each of the programs and support files has been presented in their entirety.

Typographical errors have plagued many previously published computer programs. To avoid typographical errors, each of the programs presented in the text was imported directly from the working programs. Although the programs presented here are not extremely lengthy, the taxon, element, and portion of element support files that must be created are lengthy. Anyone wishing to use this system may either create the appropriate files from the descriptions given in Appendix II, or may write to us for a copy of the programs and support files on diskette. If you are interested in receiving the programs and files on diskette, send an MS-DOS or PC-DOS preformatted 5 1/4" or 3 1/2" diskette, diskette mailer, and appropriate return postage to the authors:

Brian S. Shaffer and Barry W. Baker
FACS
Department of Anthropology
Texas A&M University
College Station, TX 77843-4352
United States

ACKNOWLEDGMENTS

The development of this coding system could not have been possible without the help of numerous persons. We thank D. Gentry Steele for his encouragement, advice, and support throughout the completion of this project. Ben W. Olive provided valuable computer programming suggestions, while W. L. McClure, Joseph F. Powell, LeeAnna Schniebs, and Bonnie C. Yates commented on earlier versions of the manuscript. We have especially enjoyed our conversations with Bonnie Yates, who has continually shared with us her zooarchaeological knowledge. Other researchers who have used this system and have provided useful comments include Becca Laws, Ben W. Olive, Julia L. Sanchez, Kristin D. Sobolik, and Laurie S. Zimmerman. We greatly appreciate these analysts for helping us test this system with their faunal assemblages. Elizabeth J. Hamm provided the cover illustration. This research was funded in part by a grant from the College of Liberal Arts, and the Department of Anthropology, Texas A&M University. Finally, we thank Marinel and everyone else who has been subjected to the phrase "coding form" for the past three years.

INTRODUCTION

Zooarchaeologists have increasingly used computers in recent years to aid in the coding, manipulation, and retrieval of data associated with faunal remains recovered from archaeological sites. With an increasing interest in zooarchaeological research, it has become important for researchers to develop a means of encoding various faunal attributes to aid site interpretations with present computer technology. Recent discussions of such computer associated zooarchaeological coding forms include Aaris-Sorensen (1981:3-29), Armitage (1978:39-45), Bayham (1982:173-177), Bonnichsen and Sanger (1977:109-133), Brumley (1973:33-36), Campana and Crabtree (1987:57-67), Clutton-Brock (1975:21-34), Cruz-Uribe and Klein (1986:171-187), Desse and Chaix (1986:93-98), Desse, Chaix, and Desse-Berset (1986)(as reviewed by Campana and Crabtree 1988:13-14), Dobney and Rielly (1988:79-96), Gifford and Crader (1977:225-238), Gifford-Gonzalez and Wright (1986:137-164), Klein and Cruz-Uribe (1984), Limp, Farley, and Andrews (1986:73-86), McArdle (1975-77:181-190), Meadow (1978:169-186), Muniz (1988:111-150), Munzel (1986:193-195; 1988:93-110), Nichol and Creak (1979:6-16), Parker and Kaczor (1986:45-71), Redding, Pires-Ferreira, and Zeder (1975-77:191-205), Redding, Zeder, and McArdle (1978:135-147), Reed (1971), Uerpmann (1978:149-167), van Wijngaarden-Bakker (1986:115-128), and Wheeler and Jones (1989:130-135).

Undoubtedly, numerous unpublished coding forms are also in existence, having been prepared by individual researchers for various research questions or projects. Two such forms were available for examination by the authors. They included a form previously developed by Jana Hellier and Cristi Assad (n.d.) of the Zooarchaeology Laboratory, Department of Anthropology, Texas A&M University, and a form developed by Bonnie C. Yates (n.d.) of the Institute of Applied Sciences at the University of North Texas. Based on insight gained from previous forms, as well as from experience, we have developed a new vertebrate faunal analysis coding form (Appendix I) and supporting dBase (Ashton-Tate 1986, 1989) software (Appendix II) that combines what we believe to be the best features of each reviewed form while providing a flexible format.

CODING FORM CRITERIA

Gifford and Crader (1977:226-227) list four prerequisites for establishing a vertebrate coding form: 1) the coding format should be intelligible, 2) the data should be sortable, 3) a letter and number coding system should be utilized that is economical in terms of time and energy, and 4) the form should be amendable and expandable. To add to this, Parker and Kaczor (1986:45) have stated that the form must be able to handle complex and dynamic data that have many interrelationships, and must allow different user views of the data.

Intelligibility may be defined as the ability to decode coded data, either visually or electronically. With current technology, this process is readily handled by simple replacement procedures that restore the coded data back into words. This process may be carried out by the computer and does not represent a major problem.

Sortability is based on the process of grouping coded data by numeric codes (Gifford and Crader 1977:226; Gifford-Gonzalez and Wright 1986:140). Such grouping depends on the ability of the computer to manipulate the data and is influenced by the form's structure, organization, and amendability. For faunal data to be sorted, the form must first be set up in an organized manner, with related elements being presented together, but numerically separated enough to allow for internal expansion (i.e. "open-ended") as noted by Bayham (1982:176) and van Wijngaarden-Bakker (1986:118). In this way, future expansion will not adversely affect sortability. This flexibility will allow the form to remain applicable through later revisions and thereby increase its useful longevity. If a form is not flexible, it diminishes its usability, sortability, and organization potential.

Gifford and Crader's (1977:227) decision to use both alphabetic and numeric entries for coding attributes does not, in our opinion, presently meet their standard of being "economical in terms of expenditure of human time and energy" (1977:227). The use of both letters and numbers, as promoted by their form, results in the coding of each attribute twice. Even Gifford and Crader (1977:228) note this problem, but suggest that alphabetic codes are easier to read and record and therefore should be included.

Gifford-Gonzalez and Wright (1986) and Klein and Cruz-Uribe (1984) reiterate the use of the alpha-numeric system promoted by Gifford and Crader (1977). However, the combining of alphabetic and numeric codes slows the initial coding process and later data entry process and is therefore perceived as being inefficient in terms of time and energy, as noted by Bayham (1982:176). Thus, we feel that coding forms should be developed based on either numeric or alphabetic characters. Numeric coding appears to be the better of the two alternatives based upon the speed of 10-key encoding. Aaris-Sorensen (1981:11) has stated that the use of a hierarchical structure in conjunction with numeric codes creates a format that is readily learned. It is based upon this concept that our coding system has been constructed.

PROPOSAL FOR A NEW DATABASE CODING SYSTEM

On the basis of Gifford and Crader's (1977) and Parker and Kaczor's (1986) criteria, the structure of a new coding system was formulated (FACS, Faunal Analysis Coding System). FACS was organized to allow for sortability, as well as expansion. Numeric codes were used to ensure economical encoding of data in terms of time and energy. This coding system was not developed simply for the purpose of constructing bone element lists, but was designed to record data such as age, sex, taxonomy, medical disorders, and taphonomic attributes. FACS was designed and tested with a commercial database management package, dBase III Plus (Ashton-Tate 1986) and dBase IV (Ashton-Tate 1989) on an IBM compatible microcomputer with 640K of random access memory, one floppy disk drive, and one 20 megabyte hard disk drive. The decision to use dBase stemmed from its wide commercial use, its user-designed format, and its potential interaction with word processing, statistical, and graphics packages. For a brief review of the advantages of dBase III Plus see Bailey, Steigman, and Peterman (1990).

The structure of the form is easily applicable to most database management software. This format was developed for use with commercial software for three reasons. First, database systems such as dBase allow for table and report generation, presenting data by the parameters established by the user and not by the programmer. This differs from zooarchaeological programs by other authors that are designed with a fixed field format (Campana and Crabtree 1987; Cruz-Uribe and Klein 1986; Klein and Cruz-Uribe 1984; McArdle 1975-77; Meadow 1978; Redding, Pires-Ferreira, and Zeder 1975-77; Redding, Zeder, and McArdle 1978; and Uerpmann 1978). Secondly, decisions as to which attributes are manipulated and how such manipulation is undertaken is determined by the user. This allows for adaptation to a greater variety of research problems. The potential for variability in manipulating data becomes very important when the numerous methods for determining MNI (minimum number of individuals) are considered, for example. Finally, by providing the essential structure for programming, this form can be adapted to different systems and is not tied to only one system or programming language. With changes in both computer technology and issues in zooarchaeological research, other forms or programs with fixed formats are more subject to becoming obsolete. Instead, FACS is designed with the flexibility to be adapted to future innovations.

FACS includes a refined list of elements for each class of vertebrate, with a total of 400 individual element codes. To compliment the element section, the portion of element section includes 322 separate codes for describing the relative completeness of an element. Different researchers have assigned varying amounts of importance to the portion of element category. McArdle (1975-77) did not include a portion of element section. Aaris-Sorensen (1981:28) included one category of 10 portion of element (fragmentation) descriptions for use with any individual element recovered. Redding, Pires-Ferreira and Zeder (1975-77), Gifford-Gonzalez and Wright (1986:149), and Hellier and Assad n.d., used a more extensive portion section, but like Aaris-Sorensen's (1981) section, they were generalized descriptors used to describe a given bone. Uerpmann (1975-77) used a detailed portion section that was subdivided by element type. In this way, bone-specific portion descriptions could be made. For Klein and Cruz-Uribe (1984) and Meadow (1978), portion of element is described in terms of fractions of the end of a long bone. For Redding, Zeder, and McArdle (1978:138), fragmentation is described in terms of both percentage and element-specific descriptions, though we feel the descriptions are not sufficient for detailed MNI analyses. Gifford and Crader (1977:237-238) and Parker and Kaczor (1986:70) chose to use two sets of coding categories to describe a given portion of an element.

We have greatly modified and expanded the portion of element section of FACS compared to those of our predecessors. Similar to Uerpmann's (1978) portion categories, but with many more portion descriptions, FACS's descriptions are divided into specific element-type portion categories (e.g. long bone portions). Since species identification can often be made based on a single tooth, an extensive tooth listing is provided. To record teeth, for example, the element would be coded as a permanent, deciduous, or indeterminate (age) tooth, with the portion of element category providing information as to the exact tooth. For long bones, we adopted a modified version of Yates' (n.d.) long bone portion classification that was derived from Munzel (1986:193-195). With the portion of element codes that we have developed, it is possible to be as specific or as general as the researcher wishes.

Since MNI is based on the most frequently recovered single element (Sellards 1952:22-23; White 1953), the portion of element category becomes critical when fragmentary bone specimens are examined. For example, the distal lateral end of an adult deer right tibia can be identified distinctly from the distal medial end. If four distal lateral ends and one distal medial end were recovered from a site, the MNI would be four. If the portion format is based on fractional values, as with previous systems (Klein and Cruz-Uribe 1984; Meadow 1978:177; Wheeler and Jones 1989:133), this determination would be impossible. Thus, the distal lateral end of a right tibia would be coded as 50% of a distal right tibia. Klein and Cruz-Uribe's (1984) computer tabulation program, "MNI," would then sum these percentages to determine an MNI. From the five tibia fragments mentioned above, their program would determine MNI to be two. This would underestimate the MNI figure. Cruz-Uribe and Klein (1986:184) have reiterated their use of this MNI calculation method in their revised PASCAL computer programs for estimating taxonomic abundance.

Klein and Cruz-Uribe's (1984:141-146) treatment of indeterminate specimens is also peculiar and may adversely affect MNI. Their "MNI" program assigns indeterminate elements with either a left or right, male or female, and fused or unfused designation. This is based on the assumption that each of these choices has a 50% chance of occurrence. There is, however, no reason to assume this to be true. Even though Klein and Cruz-Uribe (1984:144) acknowledge this statement, they feel that this will not adversely affect MNI determinations. However, MNI will be affected, and the effect may be substantial when further statistical applications are being performed with the data. Since Klein and Cruz-Uribe's "MNI" program assigns left and right, fused and unfused, and male and female designations to indeterminate bones (1984:141-146), their MNI figure becomes artificially inflated. Based on the treatment of long bone percentages that deflate MNI figures, and the assignment of additional data to indeterminate bones that artificially inflate MNI figures, the final MNI determined on a given species or site is probably more representative of the quantification methods used by Klein and Cruz-Uribe than of the number of individuals represented by the assemblage.

Though their programming represents a significant advancement in the use of computer technology in zooarchaeology, Klein and Cruz-Uribe's methods of determining MNI may not be the best choice in making such calculations. Grayson (1984:29-40) has pointed out that MNI calculations may vary widely with different aggregation techniques. Due to different research orientations to which our form may be applied, we have chosen not to suggest any one method to determine MNI. Instead, we recommend estimating MNI by visual inspection of data sorted and tabulated by category types (see XTOT in Appendix II). This is not to say that an MNI program cannot be developed for our system, but due to the number of decisions that have to be made in estimating MNI, we have found that visual inspection of sorted data may actually be quicker than computer tabulations on a personal computer.

In addition to element identification, each specimen can be coded for age and sex, as well as the criteria upon which age and sex estimations were based. Another important research issue that is included on several coding forms examined by the authors is the coverage of taphonomy (Cruz-Uribe and Klein 1986; Hellier and Assad n.d.; McArdle

1975-77; Meadow 1978; Parker and Kaczor 1986; Redding, Pires-Ferreira, and Zeder 1975-77; Uerpmann 1978; Yates n.d.). FACS allows for the coding of degree of weathering, weathering type, breakage pattern, burning, gnawing, and several forms of cultural modification.

Element Lists

The primary element list includes those elements commonly found in one or more classes of tetrapod vertebrates. This list has been developed from several standard osteological and medical sources, and from terms utilized by coding forms mentioned within this paper. Most notably, the podial list was derived from Shively (1984). In addition, we have made every effort to make the element lists as complete as possible. Many taxa have unique elements, or fused element complexes, and we have included these as well. For these unusual elements that are found in specific classes, a separate category has been provided. As examples, turtle shell elements are listed under "Reptilian Elements," Weberian vertebrae (vertebrae one through four in Ostariophysi) are listed under "Osteichthyean Elements," and the os falciforme (a supernumerary digit in the mole forelimb) is listed under "Mammalian Elements," etc.

Feather nomenclature and definitions were derived from Van Tyne and Berger (1976). Turtle shell elements were gleaned from Olsen (1968). Fish elements, listed under a separate fish element section, were compiled from a variety of sources, including Cannon (1987), Liem (1963), and Romer and Parsons (1986:184), with most of the otolith, scale, and vertebral terminology taken from Casteel (1976). Courtemanche and Legendre (1985), Rojo (1991), and Wheeler and Jones (1989) provided additional listings of osteichthyean elements, though there is some disagreement on nomenclature. Our listing applies to primarily teleostean and holostean fish. Since the most commonly preserved fish elements include otoliths, scales, and vertebrae, the provided list should be sufficient for analyzing most archaeological faunal samples. These three elements usually "provide the greatest amount of information regarding such things as identification of the fish, estimates of its size, age, growth rate, season of death, and so forth" (Casteel 1976:17). Aside from otoliths, scales, and vertebrae, two additional fish element sections are included. These provide a detailed listing of the elements of the cranial and pelvic region, and postcranial elements. For those researchers specializing in taxa for which we have not provided complete element lists, space is provided to expand the number of element codes.

Taxonomy

Some forms have included a limited listing of taxa with their attribute codes (Bonnichsen and Sanger 1977; Campana and Crabtree 1987; Desse, Chaix, and Desse-Berset 1986; Gifford-Gonzalez and Wright 1986; Hellier and Assad n.d.; McArdle 1975-77; Meadow 1978; Parker and Kaczor 1986; Redding, Zeder, and McArdle 1978; Redding, Pires-Ferreira, and Zeder 1975-77; Uerpmann 1978; and Yates n.d.). We have also chosen to include a limited listing of taxa, though greatly expanded from those of our predecessors. This list is comprised of North American taxa from north of Mexico and consists of more than 1300 taxa codes. Of the taxa presented, fish are listed to family, amphibians are listed to genus, reptiles are listed to genus with turtles being taken to species, birds are listed to genus, and mammals are listed to species. Every class was not taken to the species level for reasons of practicality; however, each taxon is numerically separated by at least 50 numbers, providing room for internal expansion. Since mammals are typically the most commonly recovered New World taxon from archaeological sites, codes for species are provided. Turtles also possess many characteristics that make species identification possible (Olsen 1968; Sobolik and Steele n.d.).

Amphibian specimens from archaeological sites are not as readily recovered as are mammalian and reptilian remains and are often undiagnostic. This is due to both morphological and structural characteristics, as well as taphonomic forces. Thus, this category has only been taken to genus. Birds have also only been listed to genus. Identification past this level is often difficult, particularly on skeletal remains. This, combined with the extremely large number of avian species present in North America and the generally fragile nature of avian elements and thus their relative scarcity in many sites, made listing by species impractical. The same is true for fish. Once more, room has been left for expansion within the taxonomic lists should more specific identifications occur.

Taxonomy for FACS has been derived from several sources. Fish taxonomy followed Nelson (1984). Amphibian and reptile taxonomy followed Stebbins (1985) and Dixon (1987). Avian taxonomy was derived from Robbins, Bruun, and Zim (1983). Mammalian taxonomy followed Hall (1981), except where certain later taxonomic changes have been introduced. Other taxonomic information was provided by Bryan, Gallucci, and Riskind (1984), Burt and Grossenheider (1976), Conant (1975), Findley (1975), Harrison and King (1965), Koster (1957), McClane (1974), Storer (1972), and Whitaker and Elman (1987).

THE CODING FORM

Once a specimen has been examined, the results can be recorded. We have chosen to make a hard copy of our work before the data is entered into the computer, though some analysts may prefer entering data directly into the computer as each specimen is identified. Two major advantages exist for first coding data onto a hard copy. First, the hard copy provides a basis for verifying that the data encoded into the computer is the data coded by the zooarchaeologist. Second, a hard copy provides a backup of the data should the data in the computer be altered or destroyed. Table 1 displays several of the categories as they would be used when coding data from a site. Data presented in this table is taken from a Classic Period Mimbres Puebloan ruin (NAN-15), located in Grant County, New Mexico (Shaffer and Miller n.d.). The use of only selected categories reflects the research orientation of the project.

Entry of the data is analogous to Gifford and Crader's (1977) procedures. As each specimen is identified, its attribute codes are entered on a given line under each attribute field. The attribute code selected for a specimen should represent the best choice of the attributes listed on the form. Once more, one or more attribute categories may be omitted or added, depending upon the research design of the project.

One characteristic of FACS that should be elaborated on is the treatment of undiagnostic or indeterminate attributes. Should a given

Table 1. Sample Data Set Displaying Codes as They Would be Entered.

Room	Qty	Taxon	Element	Portion	Side	Age	W e a t h	B r e a k	B u r n	B u t	C	Comments*
2	1	76200	210	715	2	0	1	1	0	0		
2	1	76200	30	263	1	0	1	0	0	0		
2	1	78800	20	101	1	5	1	0	0	0		
2	2	78800	30	264	1	5	1	0	0	0		
5	2	70040	140	460	0	0	1	1	0	0		
5	1	75650	442	901	1	0	1	0	0	0		
5	1	80450	210	711	1	0	1	0	0	0		
5	1	80450	20	154	1	0	1	1	0	0		
5	10	70040	0	0	0	0	1	1	0	0		
5	1	80450	30	251	2	0	1	0	0	0		
5	1	76850	170	510	2	0	1	1	0	0		
5	1	75650	240	610	2	0	1	1	0	0		
5	1	50020	190	601	0	5	1	0	0	0		
5	1	58400	810	601	1	0	1	0	0	0		
5	1	80450	10	73	1	0	1	1	0	0		
5	1	80450	30	234	2	0	1	0	0	0		
5	1	84000	240	610	2	0	1	1	0	0		
5	2	75600	90	301	3	0	1	0	0	0		
5	1	76200	220	660	0	0	1	1	0	0		
5	1	86400	30	271	2	0	1	0	0	0		
5	1	79700	30	251	1	0	1	1	0	0		
8	1	76200	220	640	2	5	1	1	0	0		
8	1	75650	180	650	1	0	1	1	1	0		
8	1	76850	20	154	2	0	1	1	0	0		
8	1	76200	190	601	2	0	1	0	0	0		
8	1	93450	90	303	3	0	1	1	0	0		
9	1	94200	442	901	2	0	1	0	0	1		

* = Field titles are expanded for readability

field not be applicable to a given element, or if the taxon or element cannot be identified, the default code is "0." This method is repeated for each category. Aaris-Sorensen (1981) and McArdle (1975-77) obviously realized the value of this method. Since indeterminate or non-applicable codes are "0," they need not be entered. Instead, the field may simply be left blank. This further speeds data coding and computer encoding. Also, each portion category has a code for complete elements and a code for indeterminate fragments. However, the codes "1" for complete elements, and "999" for unidentifiable fragments, can be used universally, regardless of the type of element being described.

Another characteristic designed to speed encoding centers around the use of the quantity (QTY) column. Instead of coding every bone on an individual line, those bones with identical attributes and provenience can be coded on the same line, their total being summed in the quantity column.

COMPUTER APPLICATIONS

FACS has been used for several faunal (Baker et al. 1991:139-161; Shaffer 1989, 1990, 1991; Shaffer and Baker 1988; Zimmerman 1990:333-355) and human analyses (Powell 1991:149-173) and subsequently modified until a concise and coherent format was established. This format has made possible the development of three support programs in dBase (Appendix II). These programs include an error detection program, a data sort program, and a program that converts the numeric codes into words for report generation.

The error detection program (CHECK) is a simple program that locates problems within and between attribute categories in several fields. This program works by flagging the line with an "error code number" that can then be used to locate and identify the error within the line much in the same way that the "Format Program" flags errors as described by Gifford-Gonzalez and Wright (1986:140). The program works by locating numbers that are too high or low for a given category. The determination of whether the number is too high or low is dependent upon the category. For example, if a radius is coded in the element category, then the portion of element category number must fall within the range of long bone portions. Otherwise, the line will be flagged. For autonomous categories such as gnawing, if the number coded is too high, then the line will be flagged with an error code.

Of special interest is the data sort program (XTOT) that has been developed for this system. This program sorts the database on as many as eight attributes defined by the user, including attributes such as provenience, taxon, element, portion of element, side, age, and sex. This sorted data is then saved to a separate file. Since MNI's may be determined in so many ways, this program has been designed to generate the relevant information upon which MNI may be determined. Here, the relevance is determined by the user through the selection of attribute fields. Once the relevant information has been sorted, the researcher can choose the methods to be used in determining MNI.

Numerical encoding of specimen information may be quicker than alpha-numeric encoding, but reading the end product is difficult without a key. Thus, a necessity for any numeric-based coding system is a decoding system for converting attribute codes back into words. This has been accomplished through a program and report form within dBase that converts coded numbers into words. Not only does this program make the coded data more readable, it also greatly aids in table construction. This program is described further in Appendix II.

Aside from the above programs, two sets of procedures have been devised to aid in data manipulation of dBase files. The first of these is a set of procedures that will condense a dBase file down to the fewest number of unique lines. These procedures are very helpful for combining coded specimens with matching attributes and proveniences into one line, with the quantity of each being totaled together on the condensed line. The second set of procedures allows the user to link two dBase files together based on a single field common to each file. In this way, faunal data may be coded based on a field sack or lot number system and then later linked to provenience information in a separate file. With this series of procedures, it is not necessary to encode all of the provenience information for each faunal specimen. A detailed description of datalink procedures are provided in Appendix II.

USE OF THE CODING FORM

We suggest that the following section be read while referring to the numeric codes themselves (Appendix I) to clarify aspects of the codes and coding process before an analysis is attempted. We realize that the extensive codes may appear overwhelming, but the listings insure the system's applicability to a wide range of assemblages. Only those sections that we feel need clarification are discussed here.

Provenience Information

The first information we record for a given faunal specimen is that of provenience. We have chosen not to list specific provenience fields or codes here due to the fact that archaeologists employ different systems in recording the horizontal and vertical location of archaeological material. As examples though, it is possible to record information such as site number, relative time period, chronometric date, unit designation, quadrant number, level, field sack number or field specimen number, and laboratory catalogue number, to name a few. We modify these fields for each analysis we conduct in conjunction with parameters established by the project director.

In recording provenience information, remember that it is possible to use a numeric or alpha-numeric code that may be later modified within the dBase file. That is, it is pointless to write out "N200W105" for every faunal specimen with this unit designation when the same information could be recorded with an alias such as "1." Actually, each unit can be designated with a different alias. As with the above example, once this code number "1" has been entered into a dBase file, it is possible to implement a dBase process that will replace all "1"'s in the provenience field with "N200W105". See Database Link Procedures in Appendix II for details describing this process.

Taxon

We have included within FACS an extensive taxonomic listing for the recording of scientific and common names. Each taxon code is numerically separated by at least 50 numbers, allowing for extensive internal expansion. It should be noted that we use a code of "00000," or simply "0," to indicate that a specimen cannot be identified past the level of Vertebrata. This speeds up the coding and data entry process because the taxon field can simply be left blank due to the fact that dBase identifies a blank field as zero. This approach has been applied to the element and portion of element fields as well. In addition to the formal taxon names, common names for each of the taxon codes are also included. Thus, in table construction, it is possible to print the formal taxon, the common name, or both. For equivocal identifications where "cf." (compares favorably) is used, we have expanded the taxon data base (not shown in Appendix I). To record a "cf." identification, the numeric code for any taxon is simply increased by one. For example, the code for *Canis familiaris* is 91850. The code 91851 therefore represents "cf. *Canis familiaris*" and will be printed as such.

Even though a specimen may not be identifiable to a formal taxonomic level beyond class, it may be possible to assign it to a size category within a given Class. In these situations, taxon codes such as small bird (50010), medium mammal (70050), or medium/large mammal (70060) may be used. While we define what we mean by these terms next to the appropriate code, researchers may wish to use their own size categories or modify our definitions.

Element

Several aspects of the element codes require explanation. These will be discussed in the order in which they appear on the form.

Common Elements

These codes refer to elements that appear in one or more vertebrate classes, with the exception of fish. We have chosen to list all fish elements separately due to the complex nature of osteichthyean osteology. Thus, in coding any fish element, code numbers 501 through 635 should be used. For example, if a fish atlas is identified in an assemblage, the element code 515 should be used. If the common element code 051 is used instead, the encoded data line will be flagged with an error code when the check program is implemented. Specific amphibian, reptilian, avian, and mammalian elements have been listed separately as well, but the majority of elements for these taxa are covered in the Common element section. The common elements are listed by axial and appendicular skeleton, respectively.

The term "cranium" (010) refers to the brain case and facial area (i.e. everything except the mandible).

In coding teeth, the element field is used to identify the tooth as permanent (030), deciduous (031), or permanent/deciduous indeterminate (032). The portion of element field is used to specify the tooth that is represented. We chose this system to avoid duplicating our tooth codes for these three categories. If the specific teeth were listed in the element field, it would have been necessary to construct codes such as Deciduous upper 1st incisor, Permanent upper 1st incisor, Upper 1st incisor deciduous/permanent indeterminate, etc. Such a system would have tripled the number of tooth codes required. It should also be noted that each tooth is coded individually, rather than describing a specimen as a "mandible/maxilla with teeth". This approach has been taken for making comparisons between loose teeth and those retained within the alveolus, and for making more accurate MNI estimations. The comments field can be used to indicate that a series of teeth is from the same individual, if needed. In such cases where teeth are retained within the maxilla or mandible, the bone should be coded separately as well.

Included with the podial codes (400 through 446) is a series of asterisks denoting mammalian groups that possess these elements. This was done in an attempt to reduce identification and coding time. We have also provided a limited listing of synonyms for the podials. The reader is referred to Peters (1987) for a more detailed discussion of alternative podial terms.

The sesamoid codes (470 through 489) have been constructed primarily for artiodactyls, though the codes are applicable to other taxonomic groups as well. Due to the difficulty involved in the siding of sesamoids, the following system is used to code these elements. *Bos taurus* (cow), for example, has four sesamoids at the junction of the metacarpal and the proximal phalanges, termed "proximal sesamoids." The proximal sesamoids consist of two axial sesamoids and two abaxial sesamoids. The proximal axial sesamoids, in this example, are located along the midline of the limb and the abaxial sesamoids are located relatively further from the center axis of the limb (Shively 1984:30).

Miscellaneous Elements

Four element codes have been included in this category that cannot be neatly placed within the Common elements. These include long bone indeterminate (490), flat bone indeterminate (492), compact bone indeterminate (494), and epiphysis indeterminate (496).

Portion of Element

The portion of element codes, like the element and taxon codes, have been constructed to allow for different levels of identification and recorder expertise. Parker and Kaczor (1986:53) formed their faunal coding system on this concept as well. For example, we have developed a very detailed series of 71 codes for recording portions of long bones (601 through 699). Some researchers may wish to be very detailed and record a long bone specimen as the "proximal posterior medial end of long bone" (617), while other researchers may be content making identifications such as "proximal end of long bone" (610), on the same specimen. Either method is acceptable, as long as the codes are used in a consistent manner. It is the need for consistency that makes it imperative that the entire form be studied and the codes reviewed before the coding process begins.

Mandible

The mandible is a difficult element to describe in terms of portions when all of the possible locations of breakage are taken into account. The mandible portions described here represent those observed during our own research, as well as portions described by Brumley (1973).

Long Bone

In the long bone portion of element section, the phrase "proximal end of long bone" refers to the proximal part of a diaphysis (shaft) and its associated fused or fusing epiphysis or articular surface. The phrase "proximal portion of long bone" refers to a diaphysis that lacks its epiphysis or articular end, either as a result of being unfused or having broken off. The same is true for the distal end and distal portion of long bone categories. The phrase "shaft of long bone" refers to a diaphysis that lacks both epiphyses or articular ends, either as a result of being unfused or having broken off.

Podials and Miscellaneous Portions of Elements

This series of codes (901 through 999) should be used to code podial portions and any element not previously covered in the portion of element section (such as scales, turtle shell elements, etc.). Again, these codes allow for different levels of identification (distal portion vs. distal posterior medial portion, for example).

Side

We have chosen to use four codes to record information on element symmetry. They include indeterminate (0), left (1), right (2), and axial (3). We decided not to use a code such as "both left and right" for problems associated with data manipulation, quantification, and MNI estimation.

Age Criteria

The majority of the age criteria codes should be self explanatory. However, a few clarifications do need to be made. In coding the fusion state of a vertebral centrum, the terms "proximal epiphysis" and "distal epiphysis" should be substituted for anterior and posterior epiphyses, respectively.

We have included four codes for recording the amount of wear observable on dentition, though we realize other researchers who work in this area use more extensive codes (Grant 1975, 1978, 1982; Levine 1982; and Payne 1973). Analysts may wish to expand the age criteria codes given here, particularly if tooth wear codes are commonly utilized. See Hillson (1986) for an excellent review of the application of dental eruption and attrition data to archaeological faunal remains.

Age

We have incorporated the use of six generalized age categories within our system. These include indeterminate (0), fetal/infant (3), subadult (5), adult (6), and old adult (8). If more specific age information is required, the codes can be expanded or a separate field can be created to include chronometric ages in months or years. The Comments field may be used for this as well.

Taphonomy

Weathering

Three codes are used here for recording the degree of mechanical or chemical reduction observed on an element. They include absent (0), slight (1), and marked (2). See Behrensmeyer (1978), Johnson (1985), and Lyman and Fox (1989) for more detailed discussions of bone weathering and alternative weathering codes.

Burning

In its present format, five codes can be used to record information on burning. These include absent (0), present (1), charred (3), calcined (7), and indeterminate (9). See Gilchrist and Mytum (1986), Shipman, Foster, and Schoeninger (1984), and Spennemann and Colley (1989) for discussions of burned bone and various colors of burned bone that have been identified. We have never undertaken an analysis of burned bone beyond the level of presence/absence, charred/calcined with FACS, though color codes could be readily incorporated. In conducting such color analyses, some standardized color classification reference should be used to provide for consistency and inter-researcher comparison. Shipman, Foster, and Schoeninger (1984) used Munsell Soil Color Charts as a standard for recording burned bone colors. Our inspection of the Munsell Soil Color charts revealed that several colors of burned bone that we have noted, especially some of the blues, were not present in the soil color charts. We suggest using more generalized color charts with a wider range of color variations, and not soil charts, for recording colors of burned bone. Ideally, a standardized light source, balanced for daylight, should be used in making color identifications, as well.

Cut Marks

Three codes are included here to record cut mark information. They include absent (0), present (1), and indeterminate (99). Assemblages that we have worked with have produced so few cut marks that we not developed a set of detailed cut mark codes. If more detailed codes are required, see Binford (1981:105-147), Grayson (1988:63), Guilday, Parmalee, and Tanner (1962), Lyman (1977, 1991:316-343), Tamplin, Haley and DeHetre (1983), and von Den Driesch and Boesneck (1975) for discussions of cut marks and cut mark codes on mammals, and Wheeler and Jones (1989:65-66) for a discussion of cut marks on fish.

Medical Disorders/Trauma

Due to the complex nature of medical disorders and traumatic injury associated with bone, we currently use only three codes for this field [absent (0), present (1), and indeterminate (9)]. If a pathological specimen is encountered, we simply record a description of it in the Comments field. See Baker and Brothwell (1980) for a discussion of the importance of identifying diseases from archaeological fauna.

Comments

The Comments field is currently 30 characters long and can be used to record any information for which a separate field has not been constructed.

SUMMARY

What we have presented within this paper are the constructs for a logical and hierarchal vertebrate coding system for use in the analysis of faunal remains from archaeological sites. FACS consists of a series of numeric codes for recording information on 24 attributes for each faunal specimen. Attribute fields include provenience, taxon, element, portion of element, side, age, sex, as well as taphonomic information. The system is designed with internal expansion capabilities, should different research questions arise. This flexible format, along with the system's adaptability to commercial software, makes it readily applicable to various projects.

In addition to the numeric coding system, a series of computer programs written for dBase software has been designed to manipulate the faunal data. These include programs for error detection, sorting, and table generation. The detection of errors is accomplished by locating data mismatches and nonexistent codes. The sort program enables the researcher to sort the database on a maximum of eight attribute fields, allowing for easy review of specific information, such as deriving counts for MNI. The table generation program converts the numeric database file into words, permitting the analyst to begin generating tables as soon as the faunal data have been entered.

APPENDIX I

VERTEBRATE FAUNAL ANALYSIS SYSTEM CODES

PROVENIENCE INFORMATION (To be determined on individual project basis)

ATTRIBUTE INFORMATION

Quantity (5)*

Taxon (5) (See page 37)
      Chondrichthyes (Cartilaginous fish) (Page 37)
      Osteichthyes (Bony fish) (Page 39)
      Amphibia (Page 47)
      Reptilia (Page 49)
      Aves (Page 53)
      Mammalia (Page 63)

Element (3) (For fish elements, see p. 21)
  Common elements (For use with tetrapod vertebrates only after 010)
  000=Indeterminate

  001=Complete skeleton

    Axial skeleton and ribs
  010=Cranium (If teeth are present, code each separately)
  020=Mandible (If teeth are present, code tooth separately)
  025=Alveolar ridge fragment (Mandible/maxilla ind.)
  030=Permanent tooth
  031=Deciduous tooth
  032=Tooth, permanent/deciduous indeterminate

  040=Vertebra indeterminate

  041=Amphicelous vertebra (Both centrum faces concave)
  042=Acelous vertebra (Mammalian, flat terminal centra faces)
  044=Procelous vertebra (Reptilian, anterior centrum concave, posterior convex)
  046=Opisthocelous vertebra (Anterior centrum convex, posterior concave)
  048=Heterocelous vertebra (Saddle-shaped articular surfaces)

  050=Cervical vertebra indeterminate
  051=Atlas
  052=Axis
  053=Cervical vertebra 3
  054=Cervical vertebra 4
  055=Cervical vertebra 5
  056=Cervical vertebra 6
  057=Cervical vertebra 7

* For database establishment, the number in parenthesis indicates the number of spaces required for the numeric code for each category.

060=Thoracic vertebra indeterminate
061=Thoracic vertebra 1
062=Thoracic vertebra 2
063=Thoracic vertebra 3
064=Thoracic vertebra 4
065=Thoracic vertebra 5
066=Thoracic vertebra 6
067=Thoracic vertebra 7
068=Thoracic vertebra 8
069=Thoracic vertebra 9
070=Thoracic vertebra 10
071=Thoracic vertebra 11
072=Thoracic vertebra 12
073=Thoracic vertebra 13
074=Thoracic vertebra 14
075=Thoracic vertebra 15
076=Thoracic vertebra 16
077=Thoracic vertebra 17
078=Thoracic vertebra 18
079=Thoracic vertebra 19
080=Thoracic vertebra 20
081=Thoracic vertebra 21

090=Lumbar vertebra indeterminate
091=Lumbar vertebra 1
092=Lumbar vertebra 2
093=Lumbar vertebra 3
094=Lumbar vertebra 4
095=Lumbar vertebra 5
096=Lumbar vertebra 6
097=Lumbar vertebra 7

100=Sacrum
110=Caudal vertebra
120=Hyoid
130=Sternum
140=Rib indeterminate
141=First rib
142=Vertebrosternal rib (True rib)
143=Vertebrochondral rib (False rib)
144=Floating rib
149=Costal cartilage

Appendicular Skeleton and Girdles
150=Clavicle
160=Coracoid
170=Scapula
180=Humerus
190=Radius
200=Ulna
210=Pelvis
220=Femur
230=Patella
240=Tibia
250=Fibula
251=Distal fibula/lateral malleolus (For portion, see
    Podial section)

260=Metapodial indeterminate
269=Metapodial of paradigit (Accessory metapodial)

270=Metacarpal indeterminate
271=Metacarpal 1
272=Metacarpal 2
273=Metacarpal 3
274=Metacarpal 4
275=Metacarpal 5
277=Fused 3rd & 4th metacarpal (Ruminants, cannon bone)
278=2nd or 4th metacarpal ("Lateral splints" in horses)
279=Metacarpal of paradigit (Accessory metacarpal)

280=Metatarsal indeterminate
281=Metatarsal 1
282=Metatarsal 2
283=Metatarsal 3
284=Metatarsal 4
285=Metatarsal 5
287=Fused 3rd & 4th metatarsal (Ruminants, cannon bone)
288=2nd or 4th metatarsal ("Lateral splints" in horses)
289=Metatarsal of paradigit (Accessory metatarsal)

290=Phalange indeterminate
299=Phalange of paradigit (Accessory phalange)

300=Proximal phalange indeterminate (First phalange)
301=Proximal phalange 1
302=Proximal phalange 2
303=Proximal phalange 3
304=Proximal phalange 4
305=Proximal phalange 5
309=Proximal phalange of paradigit

310=Middle phalange indeterminate (Second phalange)
311=Middle phalange 1
312=Middle phalange 2
313=Middle phalange 3
314=Middle phalange 4
315=Middle phalange 5
319=Middle phalange of paradigit

320=Distal phalange indeterminate (Third or terminal phalange)
321=Distal phalange 1
322=Distal phalange 2
323=Distal phalange 3
324=Distal phalange 4
325=Distal phalange 5
329=Distal phalange of paradigit

330=Front proximal phalange indeterminate (1st phalange)
331=Front proximal phalange 1
332=Front proximal phalange 2
333=Front proximal phalange 3
334=Front proximal phalange 4
335=Front proximal phalange 5

339=Front proximal phalange of paradigit

340=Front middle phalange indeterminate (2nd phalange)
341=Front middle phalange 1
342=Front middle phalange 2
343=Front middle phalange 3
344=Front middle phalange 4
345=Front middle phalange 5
349=Front middle phalange of paradigit

350=Front distal phalange indeterminate (3rd or terminal phalange)
351=Front distal phalange 1
352=Front distal phalange 2
353=Front distal phalange 3
354=Front distal phalange 4
355=Front distal phalange 5
359=Front distal phalange of paradigit

360=Hind proximal phalange indeterminate (1st phalange)
361=Hind proximal phalange 1
362=Hind proximal phalange 2
363=Hind proximal phalange 3
364=Hind proximal phalange 4
365=Hind proximal phalange 5
369=Hind proximal phalange of paradigit

370=Hind medial phalange indeterminate (2nd or non-proximal/non-distal phalange. Class Aves may have up to three medial phalanges on the hind limb)
371=Hind middle phalange 1
372=Hind middle phalange 2
373=Hind middle phalange 3
374=Hind middle phalange 4
375=Hind middle phalange 5
379=Hind middle phalange of paradigit

380=Hind distal phalange indeterminate (3rd or terminal phalange)
381=Hind distal phalange 1
382=Hind distal phalange 2
383=Hind distal phalange 3
384=Hind distal phalange 4
385=Hind distal phalange 5
389=Hind distal phalange of paradigit

390=Distal phalange sheath (Unguis)

Example animals: *=Presence	General (pig)	Carn.	Rumin.	Equid
400=Podial indeterminate				
410=Carpal indeterminate				
411=First carpal bone (Os trapezium)	*	*		*
412=Second carpal bone (Os trapezoideum)	*	*		*
413=Third carpal bone (Magnum or os capitatum)	*	*		*
414=Fourth carpal bone (Unciform or os hamatum)	*	*		*
415=Fifth carpal bone (Rarely present)				*
418=Fused second & third carpal (Trapezoid-magnum)			*	
420=Radial carpal bone (Os scaphoideum)	*		*	*
422=Intermediate carpal bone (Os lunatum)	*		*	*
424=Fused radial & intermediate (Scapho-lunar)		*		
426=Ulnar carpal bone (Cuneiforme, os triquetrum)	*	*	*	*
428=Accessory carpal bone (Os pisiforme)	*	*	*	*

	General/carn.	Rumin.	Equid
430=Tarsal indeterminate			
431=First tarsal (Os cuneiforme mediale)	*	*	
432=Second tarsal	*		
433=Third tarsal (Os cuneiforme laterale)	*		*
434=Fourth tarsal (Os cuboideum)	*		*
436=Fused first & second tarsal (Os cuneiforme intermedium)			*
438=Fused second & third tarsal		*	
440=Astragalus (Talus)	*	*	*
442=Calcaneus	*	*	*
444=Central tarsal bone (Os naviculare)			*
446=Fused central tarsal & fourth tarsal (Naviculo-cuboid)		*	

XXX=Lateral malleolus--see Distal Fibula, #251, page 17.

470=Sesamoid indeterminate

471=Proximal sesamoid
472=Proximal axial sesamoid
473=Proximal abaxial sesamoid
474=Proximal sesamoid of forelimb
475=Proximal axial sesamoid of forelimb
476=Proximal abaxial sesamoid of forelimb
477=Proximal sesamoid of hind limb

478=Proximal axial sesamoid of hind limb
479=Proximal abaxial sesamoid of hind limb

481=Distal sesamoid
482=Distal axial sesamoid
483=Distal abaxial sesamoid
484=Distal sesamoid of forelimb
485=Distal axial sesamoid of forelimb
486=Distal abaxial sesamoid of forelimb
487=Distal sesamoid of hind limb
488=Distal axial sesamoid of hind limb
489=Distal abaxial sesamoid of hind limb

Miscellaneous Elements
490=Long bone indeterminate
492=Flat bone indeterminate (Ribs, scapulae, pelvis, cranium, etc.)
494=Compact bone indeterminate (Podials, sesamoids, patellae, etc.)
496=Epiphysis indeterminate

Osteichthyean/Chondrichthyean elements
Otoliths
501=Otolith indeterminate
502=Asteriscus otolith
503=Lapillus (Utricular) otolith
504=Sagitta otolith

Scales and scutes
505=Scale indeterminate
506=Placoid scale (Keeled shape--sharks, rays, and relatives)
507=Ganoid scale (Rhombic shape--gars, sturgeons, and paddlefish)
508=Cycloid scale (Disc shape--both fresh & marine fish)
509=Ctenoid scale (Disc shape w/ctenii--both fresh and marine fish)
510=Scute indeterminate
511=Dorsal scute
512=Ventral scute
513=Lateral scute

Vertebral column and spines
514=Proatlas (Vertebral face on posterior end of the basioccipital)
515=Atlas
516=Second vertebra (Posterior to atlas)
517=Weberian vertebra (Vertebrae 1-4 in Ostariophysi)
518=Thoracic vertebra
519=Precaudal vertebra (Possess fused neural spine & has developed transverse processes)
520=Caudal vertebra
521=Penultimate vertebra (Between caudal and ultimate vertebra)
522=Ultimate vertebra (Last vertebra in the series)
523=Cyclospondyl vertebra (Chondrichthyean fish)
524=Asterospondyl vertebra (Chondrichthyean fish)

525=Tectispondyl vertebra (Chondrichthyean fish)
526=Fish vertebra indeterminate
527=Spine indeterminate
528=Pectoral spine
529=Dorsal spine
530=Neural spine
531=Haemal spine
532=Pelvic spine

Elements of the cranial and pelvic region
534=Tooth
535=Adnasal
536=Alisphenoid
537=Angular
538=Articular
539=Basibranchial
540=Basibranchial plate
541=Basihyal
542=Basioccipital
543=Basipterygium
544=Branchiostegal ray
545=Ceratobranchial
546=Ceratohyal
547=Clavicular
548=Cleithrum
549=Coracoid
550=Dental bearing element
551=Dentary
552=Ectopterygoid
553=Entopterygoid
554=Epibranchial
555=Epihyal
556=Epiotic
557=Ethmoid
558=Ethmonasal
559=Exoccipital
560=Frontal
561=Gular
562=Hyomandibular
563=Hypobranchial
564=Hypohyal
565=Interhaemal spine
566=Interhyal
567=Interoperculum
568=Lacrimal
569=Maxilla
570=Mesocoracoid
571=Mesopterygoid
572=Metapterygoid
573=Nasal
574=Operculum
575=Opisthotic
576=Orbitosphenoid
577=Palatine
578=Parasphenoid
579=Parietal
580=Pharyngobranchial

581=Pharyngeal plate
582=Postclavicle
583=Postcleithrum
584=Postfrontal
585=Postorbital
586=Postparietal
587=Postemporal
588=Prefrontal
589=Premaxilla
590=Prootic
591=Preoperculum
592=Ptergiophores (Proximal radiacia)
593=Pterotic
594=Pterygials
595=Quadrate
596=Radial
597=Retroarticular
598=Rostral
599=Scapula
600=Sphenotic
601=Suboperculum
602=Suborbital
603=Suprangular
604=Supracleithrum
605=Supramaxillary
606=Supraoccipital
607=Supraorbital
608=Suprapreopercular
609=Supratemporal
610=Symplectic
611=Tabular
612=Urohyal
613=Vomer

Postcranial elements other than vertebrae
620=Epipleural
625=Hypural bone
630=Rib
635=Subperitoneal (False rib)

Amphibian elements (For non class-specific elements, see page 16)
651=Tympanic ring
653=Hyoid plate
660=Suprascapula
665=Epicoracoid
670=Episternum
671=Omosternum
672=Mesosternum
673=Xiphisternum
680=Radioulna
685=Tibiofibula
690=Urostyle

Reptilian elements (For non class-specific elements, see page 16)
701=Shell fragment indeterminate
702=Carapace fragment indeterminate
703=Nuchal
704=Neural
705=Pleural
706=Peripheral
707=Suprapygal
708=Pygal
720=Plastron fragment indeterminate
721=Epiplastron
722=Entoplastron
723=Hyoplastron
724=Hypoplastron
726=Xiphiplastron
730=Eye ring
731=Os palpebrae (Eyelid plate of crocodilians)
740=Dorsal vertebra (All vertebra with rib articular facets)
750=Dermal scale (Other than turtle shell)
751=Epidermal scale (Also includes turtle scutes)

Avian elements (For non class-specific elements, see page 16)
801=Beak, upper mandible
802=Beak, lower mandible
803=Upper beak cover/sheath
804=Lower beak cover/sheath
805=Quadrate
806=Tracheal cartilage (Ring)
808=Eye ring
810=Carpometacarpus
815=Tibiotarsus
820=Tarsometatarsus
825=Sternal rib
830=Lumbosacrale (Lumbar vertebra fused with sacrum)
835=Fused thoracic vertebrae
XXX=Synsacrum (Code as pelvis element, synsacrum portion of element)
840=Pygostyle (Fused caudal vertebrae)
845=Furculum (Wish bone)
850=Epidermal scale

855=Feather indeterminate
861=Contour feather indeterminate
862=Remige feather (Contour flight feather ind.)
863=Primary feather (Contour feather of the manus)
864=Secondary feather (Contour feather of the ulna)
865=Retrice feather (Tail feather indeterminate)
866=Semiplume feather (Loose webbed contour feather)
870=Filoplume feather (Specialized hair-like feather)
875=Bristle feather
880=Down feather
885=Powder down feather

Mammalian elements (For non class-specific elements, see page 16)
901=Antler
910=Horn
915=Malleus
916=Incus
917=Stapes
920=Os penis (Baculum)
930=Os clitoris
932=Os rosti (Rostral bone)
934=Os cordis (Artiodactyl heart bone)
936=Os falciforme (Supernumerary digit in mole forelimb)
940=Epipubic bone (Marsupials)
950=Dermal armor plate (Carapace) (Armadillos)

960=Quill (Porcupine)
970=Hair
972=Guard hair (Capilli)
974=Wool hair (Pili lanei)
976=Bristle hair
978=Long hair (Equines)
980=Tactile hair

Portion of Element (3)
  000=Indeterminate

  Cranium
  001=Complete or nearly complete

  003=Calvaria (Brain case)
  006=Frontal (If antler pedicle is present, code under
      Pedicle/Antler/Horn portion category instead of
      Skull category)
  007=Frontal fragment
  010=Parietal
  013=Interparietal
  016=Temporal
  018=Auditory bulla
  020=Petrosal (Portion of the temporal)
  023=Occiput
  026=Occipital
  030=Occipital condyle
  033=Exoccipital
  036=Basioccipital
  040=Supraoccipital
  043=Presphenoid
  046=Sphenoid
  050=Basisphenoid
  053=Alisphenoid
  054=Orbitospheniod
  056=Ethmoid
  060=Pterygoid

  063=Facial area
  066=Nasal
  070=Vomer
  073=Premaxilla (If teeth are present, code each tooth
      separately)
  076=Maxilla (If teeth are present, code each tooth
      separately)
  080=Alveolar ridge fragment
  083=Palatine
  086=Lacrimal
  090=Jugal
  093=Zygomatic
  096=Zygomatic process

  099=Cranial fragment indeterminate

  Mandible (If teeth are present, code each tooth
      separately)
  101=Complete or nearly complete (Both halves)
  111=Ramus complete or nearly complete (One half of
      mandible)
  121=Ascending ramus complete with angular process,
      coronoid process, and mandibular condyle
  124=Ascending ramus with portions of angular process and
      mandibular condyle (Coronoid process not present)
  125=Ascending ramus with mandibular condyle and
      coronoid process (Other areas not present)

127=Ascending ramus with mandibular condyle (Other areas not present)
130=Ascending ramus with coronoid process (Other areas not present)
133=Ascending ramus with angular process (Other areas not present)
135=Ascending ramus with portion of horizontal ramus
136=Portions of mandibular condyle and coronoid process
139=Angular process
142=Coronoid process
145=Mandibular condyle
148=Ascending ramus fragment indeterminate
151=Horizontal ramus with incisor alveolus
154=Horizontal ramus portion (Area or portion of area of cheekteeth)
157=Horizontal ramus with diastema (Incisor area broken off)
160=Horizontal ramus diastema
163=Horizontal ramus, incisor and diastema area only
169=Horizontal ramus fragment indeterminate
172=Alveolar ridge fragment of cheekteeth
175=Alveolar ridge fragment of incisors
179=Alveolar ridge fragment indeterminate
199=Mandibular fragment indeterminate

Tooth (If maxilla or mandible present, code separately)
201=Tooth indeterminate (Complete or nearly complete)
202=Tooth fragment (With part of enamel & part of root)
203=Enamel fragment
204=Root fragment
208=Incisor indeterminate (Upper/lower indeterminate)
209=Canine indeterminate (Upper/lower indeterminate)
210=Premolar indeterminate (Upper/lower indeterminate)
211=Molar indeterminate (Upper/lower indeterminate)
214=Upper cheek tooth (PM or M) indeterminate
215=Lower cheek tooth (PM or M) indeterminate
216=Cheek tooth indeterminate (Upper/lower, PM/M indeterminate)
217=Supernumerary tooth (Upper/lower ind., describe in Comments)
218=Supernumerary upper tooth (Describe in Comments)
219=Supernumerary lower tooth (Describe in Comments)

220=Upper I indeterminate
221=Upper I1
222=Upper I2
223=Upper I3
224=Upper I4
225=Upper I5

228=Upper C

230=Upper PM indeterminate
231=Upper PM1
232=Upper PM2
233=Upper PM3
234=Upper PM4

235=Upper PM3 or 4

238=Upper PM4 or M1 indeterminate

240=Upper M indeterminate
241=Upper M1
242=Upper M2
243=Upper M3
244=Upper M4
245=Upper M1 or 2
246=Upper M2 or 3
247=Upper M3 or 4

250=Lower I indeterminate
251=Lower I1
252=Lower I2
253=Lower I3
254=Lower I4

258=Lower C

260=Lower PM indeterminate
261=Lower PM1
262=Lower PM2
263=Lower PM3
264=Lower PM4
265=Lower PM3 or 4

268=Lower PM4 or M1 indeterminate

270=Lower M indeterminate
271=Lower M1
272=Lower M2
273=Lower M3
274=Lower M4
275=Lower M1 or 2
276=Lower M2 or 3
277=Lower M3 or 4

Vertebral column
301=Complete or nearly complete
303=Complete except spinous process not present
306=Centrum and neural area (With one or more processes not present)
309=Neural area only (Arch structure)
312=Centrum
315=Centrum epiphysis indeterminate
318=Anterior centrum epiphysis
321=Posterior centrum epiphysis
324=Odontoid process (Dens)
327=Spinous process
330=Transverse process
333=Fragment of process
336=Articular facet
339=Anterior zygapophysis
342=Posterior zygapophysis
350=Sagittal split, left portion

352=Sagittal split, right portion
354=Transverse split, cranial portion
356=Transverse split, caudal portion

360=Sacral element indeterminate
361=Sacral element 1
362=Sacral element 2
363=Sacral element 3
364=Sacral element 4
365=Sacral element 5
366=Sacral element 6
369=Sacral centrum indeterminate
370=Sacral wing (Lateral mass)
373=Sacral body
376=Sacral apex
390=Sacrum fragment indeterminate

399=Vertebra fragment indeterminate

Sternum
401=Complete or nearly complete
404=Manubrium
408=Sternal body
410=Sternal body segment
412=Xiphoid process

    Bird sternum portions (Code as 401 if
    complete/nearly complete)
420=Manubrium sterni
422=Anterior portion with manubrium sterni, coracoid
    facets, & keel
424=Crista sterni fragment (Keel)
426=Apex cristae fragment (Cranial portion of keel)
428=Metasternum fragment
430=Costosternal facets
432=Sterno-coracoidal process
434=Lateral caudal process

449=Sternal fragment indeterminate

Rib
451=Complete or nearly complete
455=Vertebral end (Head, neck, and tubercle)
460=Shaft fragment
465=Rib epiphysis
470=Sternal end and portion of shaft

Scapula
501=Complete or nearly complete
505=Glenoid fossa only
510=Glenoid fossa and incomplete blade
515=Blade portion
520=Spine fragment
525=Neck
530=Acromion process
535=Coracoid process
540=Epiphysis of scapula

599=Scapula fragment indeterminate

Long bone (Also includes metapodials and phalanges.
Note: Miscellaneous long bone portions
begin with code 680)
601=Complete or nearly complete (If no epiphyses or
   articulation present, see 640)
602=Complete minus proximal epiphysis
603=Complete minus distal epiphysis
605=Complete minus lateral portion
606=Complete minus medial portion
607=Complete minus anterior portion
508=Complete minus posterior portion

Proximal end of long bone ("End" = articulation &
   part of shaft)
610=Proximal end of long bone
612=Proximal anterior end (Both lateral and medial)
613=Proximal anterior lateral end of long bone
614=Proximal anterior medial end of long bone
615=Proximal posterior end (Both lateral and medial)
616=Proximal posterior lateral end of long bone
617=Proximal posterior medial end of long bone
618=Proximal lateral end of long bone
619=Proximal medial end of long bone

Proximal epiphysis only (No shaft fragment present)
621=Proximal epiphysis, complete or nearly complete
622=Proximal epiphysis, anterior aspect (Both lateral
   and medial)
623=Proximal epiphysis, anterior lateral aspect
624=Proximal epiphysis, anterior medial aspect
625=Proximal epiphysis, posterior aspect (Both lateral
   and medial)
626=Proximal epiphysis, posterior lateral aspect
627=Proximal epiphysis, posterior medial aspect
628=Proximal epiphysis, lateral aspect
629=Proximal epiphysis, medial aspect

Proximal portions of long bone (No articular end
   present)
630=Proximal portion of shaft
631=Proximal anterior portion of shaft (Both lateral
   and medial)
632=Proximal anterior lateral portion of shaft
633=Proximal anterior medial portion of shaft
634=Proximal posterior portion of shaft (Both lateral
   and medial)
635=Proximal posterior lateral portion of shaft
636=Proximal posterior medial portion of shaft
637=Proximal lateral portion of shaft
638=Proximal medial portion of shaft
639=Semi-lunar notch of ulna only

Shaft of long bone (No articular ends present)
640=Complete shaft (No epiphyses)
641=Lateral portion of shaft

642=Medial portion of shaft
643=Anterior portion of shaft
644=Posterior portion of shaft
645=Fibular scar on tibia

    Distal portions of long bone (No articular end present)
650=Distal portion of shaft
651=Distal anterior portion of shaft (Both lateral and medial)
652=Distal anterior lateral portion of shaft
653=Distal anterior medial portion of shaft
654=Distal posterior portion of shaft (Both lateral and medial)
655=Distal posterior lateral portion of shaft
656=Distal posterior medial portion of shaft
657=Distal lateral portion of shaft
658=Distal medial portion of shaft

    Distal end of long bone ("End" = articulation & part of shaft)
660=Distal end of long bone
662=Distal anterior end (Both lateral and medial)
663=Distal anterior lateral end of long bone
664=Distal anterior medial end of long bone
665=Distal posterior end (Both lateral and medial)
666=Distal posterior lateral end of long bone
667=Distal posterior medial end of long bone
668=Distal lateral end of long bone
669=Distal medial end of long bone

    Distal ephipysis only (No shaft fragment present)
671=Distal epiphysis, complete or nearly complete
672=Distal epiphysis, anterior aspect (Both lateral and medial)
673=Distal epiphysis, anterior lateral aspect
674=Distal epiphysis, anterior medial aspect
675=Distal epiphysis, posterior aspect (Both lateral and medial)
676=Distal epiphysis, posterior lateral aspect
677=Distal epiphysis, posterior medial aspect
678=Distal epiphysis, lateral aspect
679=Distal epiphysis, medial aspect

680=Distal articular condyle (Lateral/medial indeterminate)
681=Diaphyseal fragment
699=Long bone fragment indeterminate

Pelvis
701=Complete or nearly complete (Both halves), with synsacrum in birds
711=Os coxa complete or nearly complete (Innominate)
715=Acetabulum complete with portions of ilium, ischium, & pubis
720=Acetabulum with portion of ilium and ischium
725=Acetabulum with portion of ilium and pubis

730=Acetabulum with portion of ischium and pubis
741=Ilium complete
742=Iliac crest
743=Iliac epiphysis
745=Acetabular end of ilium
749=Ilium fragment
751=Ischium complete
755=Acetabular end of ischium
757=Pubic end of ischium
759=Ischium fragment
761=Pubis complete
765=Acetabular end of pubis
767=Ischial end of pubis
769=Pubis fragment

    Avian pelvis
790=Synsacrum (If one os coxa present, code separately)

799=Pelvis fragment indeterminate

Pedicle/Antler/Horn
801=Complete antler
803=Tip of tine
805=Antler tine (Any branch)
807=Antler fork
809=Beam, antler shed
811=Beam, antler removed culturally
813=Beam, antler detachment indeterminate
815=Pedicle, antler shed
817=Pedicle, antler removed culturally
819=Pedicle, detachment indeterminate
821=Pedicle and beam portion
829=Antler fragment indeterminate

841=Horn core complete or nearly complete
843=Tip of horn core
849=Horn core fragment
851=Horn sheath complete or nearly complete
859=Horn sheath fragment

Podials and Miscellaneous portions of elements
    (Elements not otherwise covered such as scales,
    turtle shell, podials, etc.)
901=Complete or nearly complete
902=Complete minus proximal epiphysis
903=Complete minus distal epiphysis
910=Proximal aspect
911=Proximal medial aspect
912=Proximal lateral aspect
913=Proximal anterior aspect
914=Proximal anterior medial aspect
915=Proximal anterior lateral aspect
916=Proximal posterior aspect
917=Proximal posterior medial aspect
918=Proximal posterior lateral aspect
919=Proximal aspect minus proximal epiphysis
920=Proximal epiphysis

930=Distal aspect
931=Distal medial aspect
932=Distal lateral aspect
933=Distal anterior aspect
934=Distal anterior medial aspect
935=Distal anterior lateral aspect
936=Distal posterior aspect
937=Distal posterior medial aspect
938=Distal posterior lateral aspect
939=Distal aspect minus distal epiphysis
940=Distal epiphysis
941=Medial aspect
945=Medial anterior aspect
949=Medial posterior aspect
951=Lateral aspect
955=Lateral anterior aspect
959=Lateral posterior aspect
961=Anterior aspect
971=Posterior aspect
999=Fragment

Side (1)
0=Indeterminate
1=Left
2=Right
3=Axial

Age Criteria (2)
00=Non-applicable or determination cannot be made
01=Epiphysis fused
02=Epiphysis fusing
03=Epiphysis unfused
04=Proximal fused, distal fused (For vertebrae, proximal = anterior, distal = posterior)
05=Proximal fused, distal fusing
06=Proximal fused, distal unfused
07=Proximal fusing, distal fused
08=Proximal fusing, distal fusing
09=Proximal fusing, distal unfused
10=Proximal unfused, distal fused
11=Proximal unfused, distal fusing
12=Proximal unfused, distal unfused
15=Cranial sutures fused
16=Cranial sutures unfused
20=Tooth bud (Root unformed)
22=Erupting tooth
24=Open tooth root
31=Slight or no wear on tooth
32=Moderate wear on tooth
33=Marked wear on tooth
34=Very heavy wear (Worn to root)
35=Resorption of alveolus
36=Deciduous tooth
37=Permanent tooth
38=Dental arcade with both deciduous and permanent teeth
40=Bone texture and/or size
50=Measurement

    55=Antler/horn development
    60=Arthritic lipping
    65=Incremental structures (Describe in Comments)

Age (1)
    0=Indeterminate
    3=Fetal/infant
    5=Subadult
    6=Adult
    8=Old adult

Sex Criteria (1)
    0=Non-applicable or determination cannot be made
    1=Antler or antler pedicle present/absent
    2=Os penis (Baculum)
    3=Os clitoris
    4=Metatarsal spur (Calcar) (Some male birds)
    5=Bone measurement
    6=Tooth measurement
    7=Qualitative morphology
    8=Medullary bone (In females of some species)

Sex (1)
    0=Indeterminate
    1=Male
    2=Female

TAPHONOMY

Weathering (1) (Mechanical and/or chemical reduction)
    0=Absent
    1=Slight
    2=Marked

Forms of reduction (2) (Weathering types)
    00=Absent
    01=Fine line fractures
    02=Spalling (Flaking in planes)
    03=Root etching
    04=Chemical etching
    05=Abrasion (Impact of wind and/or waterborne particles)
    06=Pitting
    07=Fine line fractures and spalling
    08=Water wear
    98=Indeterminate form of reduction
    99=Multiple forms not already covered (List in Comments)

Breakage Pattern (1)
    0=Unbroken
    1=Angular fracture (Dry bone)
    2=Spiral fracture (Green bone)
    3=Both angular and spiral fractures
    4=Cut/sawed only, no break
    5=Cut/sawed with angular fracture
    6=Cut/sawed with spiral fracture
    7=Cut/sawed with both angular and spiral fractures

Burning (1)
    0=Absent
    1=Present
    3=Charred (Black)
    7=Calcined (White)
    9=Indeterminate (Possible burning)

Gnawing (1)
    0=Absent
    1=Present
    2=Rodent
    3=Carnivore
    4=Artiodactyl
    5=Human
    6=Rodent and carnivore
    7=Gnawing, form unknown
    8=Multiple forms not already listed (List in Comments)
    9=Indeterminate (Possible gnawing)

Cut marks (2)
    00=Absent
    01=Present
    99=Indeterminate (Possible cut marks)

Drilled (1)
    0=Absent
    1=Present
    9=Indeterminate (Possibly drilled)

Polished (1)
    0=Absent
    1=Present
    9=Indeterminate (Possibly polished)

Sawed (1)
    0=Absent
    1=Present
    9=Indeterminate (Possibly sawed)

Grooved/Incised (1)
    0=Absent
    1=Present
    9=Indeterminate (Possibly grooved or incised)

Ground (1)
    0=Absent
    1=Present
    9=Indeterminate (Possibly ground)

Excavation/lab damage (1)
    0=Absent
    1=Present
    9=Indeterminate (Possible excavation or lab damage)

Other Modification (1) (Not listed above)
   0=Absent
   1=Present (Describe in Comments)
   9=Indeterminate (Possible modification)

Medical disorders/Trauma (2)
   0=Absent
   1=Present (Describe in Comments)
   9=Indeterminate (Possible pathology/trauma)

Comments (30)

TAXON (5) (Primarily United States Holocene Fauna)

    00000=Vertebrata, class indeterminate
    00010=Micro vertebrate, class indeterminate
    00020=Small vertebrate, class indeterminate
    00030=Medium vertebrate, class indeterminate
    00040=Large vertebrate, class indeterminate
    00050=Very large vertebrate, class indeterminate

Class Chondrichthyes (Cartilaginous fish--shark, rays, and sawfish)
    00100=Chondrichthyes order indeterminate

Order Chimaeriformes (Chimaeras)
    00200=Chimaeridae (Chimaeras or ratfish)

Order Hexanchiformes (Sharks)
    00300=Hexanchidae (Cow sharks)

Order Lamniformes (Sharks)
    00400=Lamniformes family indeterminate

  Family Orectolobidae (Nurse sharks)
    00500=Orectolobidae genus indeterminate

  Family Rhincodontidae (Whale sharks)
    00600=Rhincodontidae genus indeterminate

  Family Odontaspididae (Sand tiger sharks)
    00700=Odontaspididae genus indeterminate

  Family Lamnidae (Thrasher, basking, mackerel sharks)
    00800=Lamnidae genus indeterminate

  Family Carcharhinidae (Requiem sharks)
    00900=Carcharhinidae genus indeterminate

  Family Sphyrnidae (Hammerhead sharks)
    01000=Sphyrnidae genus indeterminate

Order Squaliformes (Sharks)
    01100=Squaliformes family indeterminate

  Family Squalidae (Spiney dogfish sharks)
    01200=Squalidae genus indeterminate

  Family Squatinidae (Angel sharks)
    01300=Squatinidae genus indeterminate

Order Rajiformes (Sawfish and rays)
    01400=Rajiformes family indeterminate (Sawfish and rays)

  Family Pristidae (Sawfish)
    01500=Pristidae genus indeterminate

Family Rhinobatidae (Guitarfish)
  01600=Rhinobatidae genus indeterminate

Family Torpedinidae (Electric rays)
  01700=Torpedinidae genus indeterminate

Family Rajidae (Skates)
  01800=Rajidae genus indeterminate

Family Dasyatidae (Stingrays and butterfly rays)
  01900=Dasyatidae genus indeterminate

Family Myliobatididae (Eagle rays)
  02000=Myliobatididae genus indeterminate

Family Mobulidae (Manta rays)
  02100=Mobulidae genus indeterminate

Class **Osteichthyes** (Bony fish)
    10010=Small fish indeterminate
    10030=Medium fish indeterminate
    10050=Large fish indeterminate
    10100=Osteichthyes order indeterminate

Order **Acipenseriformes** (Paddlefish and sturgeons)
    10200=Acipenseriformes family indeterminate

  Family Polyodontidae (Paddlefish)
    10250=Polyodontidae genus indeterminate

  Family Acipenseridae (Sturgeons)
    10300=Acipenseridae genus indeterminate

Order **Lepisosteiformes** (Gars)
  Family Lepisosteidae (Gars)
    10350=Lepisosteidae genus indeterminate

Order **Amiiformes** (Bowfins)
  Family Amiidae (Bowfins)
    10400=Amiidae genus indeterminate

Order **Osteoglossiformes** (Mooneyes, goldeneyes, etc.)
  Family Hiodontidae (Mooneyes and goldeneyes)
    10450=Hiodontidae genus indeterminate

Order **Elopiformes** (Tenpounders, tarpons, and bonefish)
    10500=Elopiformes family indeterminate

  Family Elopidae (Tenpounders)
    10550=Elopidae genus indeterminate

  Family Megalopidae (Tarpons)
    10600=Megalopidae genus indeterminate

  Family Albulidae (Bonefish)
    10650=Albulidae genus indeterminate

Order **Anguilliformes** (Eels)
    10700=Anguilliformes family indeterminate

  Family Anguillidae (Freshwater eels)
    10750=Anguillidae genus indeterminate

  Family Muraenidae (Moray eels)
    10800=Muraenidae genus indeterminate

  Family Congridae (Conger eels)
    10850=Congridae genus indeterminate

  Family Ophichthidae (Snake and worm eels)
    10900=Ophichthidae genus indeterminate

  Family Synaphobranchidae (Cutthroat eels)
    10950=Synaphobranchidae genus indeterminate

Order Clupeiformes (Gizzardshad, Menhadens, and Anchovies)
    11000=Clupeiformes family indeterminate

  Family Clupeidae (Menhadens)
    11050=Clupeidae genus indeterminate

  Family Engraulididae (Anchovies)
    11100=Engraulididae genus indeterminate

Order Cypriniformes (Suckers, minnows, etc.)
    11150=Cypriniformes family indeterminate

  Family Catostomidae (Suckers)
    11200=Catostomidae genus indeterminate

  Family Cyprinidae (Minnows and carp)
    11250=Cyprinidae genus indeterminate

Order Characiformes (Characids)
  Family Characidae (Characids)
    11300=Characidae genus indeterminate

Order Siluriformes (Catfish)
    11350=Siluriformes family indeterminate

  Family Ictaluridae (Catfish)
    11400=Ictaluridae genus indeterminate

  Family Ariidae (Marine catfish)
    11450=Ariidae genus indeterminate

Order Salmoniformes (Salmon, trout, etc.)
    11500=Salmoniformes family indeterminate

  Family Salmonidae (Salmon and trout)
    11550=Salmonidae genus indeterminate

  Family Esocidae (Pikes and pickerels)
    11600=Esocidae genus indeterminate

  Family Osmeridae (Smelts)
    11650=Osmeridae genus indeterminate

Order Aulopiformes (Lizardfish)
  Family Synodontidae (Lizardfish)
    11700=Synodontidae genus indeterminate

Order Percopsiformes (Pirate and troutperch)
  Family Aphredoderidae (Pirate perch)
    11750=Aphredoderidae genus indeterminate

Order Gadiformes (Codlets and codfish)
    11800=Gadiformes family indeterminate

  Family Bregmacerotidae (Codlets)
    11850=Bregmacerotidae genus indeterminate

Family Gadidae (Codfish)
  11900=Gadidae genus indeterminate

Family Amblyopsidae (Cavefish)
  11950=Amblyopsidae genus indeterminate

Order Ophidiiformes (Cusk eels and Pearfish)
  12000=Ophidiiformes family indeterminate

Family Ophidiidae (Cusk eels)
  12050=Ophidiidae genus indeterminate

Family Carapidae (Pearfish)
  12100=Carapidae genus indeterminate

Order Batrachoidiformes (Toadfish)
Family Batrachoididae (Toadfish)
  12150=Batrachoididae genus indeterminate

Order Lophiiformes (Frogfish and batfish)
  12200=Lophiiformes family indeterminate

Family Lophiidae (Goosefish)
  12250=Lophiidae genus indeterminate

Family Antennariidae (Frogfish)
  12300=Antennariidae genus indeterminate

Family Ogcocephalidae (Batfish)
  12350=Ogcocephalidae genus indeterminate

Order Gobiesociformes (Clingfish)
Family Gobiesocidae (Clingfish)
  12400=Gobiesocidae genus indeterminate

Order Cyprinodontiformes (Killifish and livebearers)
  12450=Cyprinodontiformes family indeterminate

Family Cyprinodontidae (Killifish)
  12500=Cyprinodontidae genus indeterminate

Family Poeciliidae (Livebearers)
  12550=Poeciliidae genus indeterminate

Family Belonidae (Needlefish)
  12600=Belonidae genus indeterminate

Family Exocoetidae (Flyingfish and halfbeaks)
  12650=Exocoetidae genus indeterminate

Family Scomberesocidae (Sauries)
  12700=Scomberesocidae genus indeterminate

Order Atheriniformes (Silversides)
Family Atherinidae (Silversides)
  12750=Atherinidae genus indeterminate

Order Beryciformes (Squirrelfish)
  Family Holocentridae (Squirrelfish)
    12800=Holocentridae genus indeterminate

Order Zeiformes
  Family Zeidae (Dories)
    12850=Zeidae genus indeterminate

Order Syngnathiformes (Pipefish, etc.)
    12900=Syngnathiformes family indeterminate

  Family Syngnathidae (Pipefish and seahorses)
    12950=Syngnathidae genus indeterminate

  Family Aulostomidae (Trumpeter fish)
    13000=Aulostomidae genus indeterminate

  Family Fistulariidae (Cornet fish)
    13050=Fistulariidae genus indeterminate

Order Dactylopteriformes (Flying gurnards)
  Family Dactylopteridae (Flying gurnards)
    13100=Dactylopteridae genus indeterminate

Order Scorpaeniformes (Scorpionfish and sea robins)
    13150=Scorpaeniformes family indeterminate

  Family Scorpaenidae (Scorpionfish)
    13200=Scorpaenidae genus indeterminate

  Family Agonidae (Poachers)
    13250=Agonidae genus indeterminate

  Family Triglidae (Sea robins)
    13300=Triglidae genus indeterminate

  Family Hexagrammidae (Greenlings)
    13350=Hexagrammidae genus indeterminate

  Family Cyclopteridae (Lumpfish and snailfish)
    13400=Cyclopteridae genus indeterminate

Order Perciformes (Perches, basses, etc.)
    13450=Perciformes family indeterminate

  Family Centropomidae (Snooks)
    13500=Centropomidae genus indeterminate

  Family Serranidae (Sea basses)
    13550=Serranidae genus indeterminate

  Family Centrarchidae (Sunfish, blackbass, and crappies)
    13600=Centrarchidae genus indeterminate

  Family Percidae (Perches, pikeperches, and darters)
    13650=Percidae genus indeterminate

Family Carangidae (Jacks)
  13700=Carangidae genus indeterminate

Family Sciaenidae (Drums, croakers, etc.)
  13750=Sciaenidae genus indeterminate

Family Sparidae (Porgies and sheepsheads)
  13800=Sparidae genus indeterminate

Family Cichlidae (Cichlids)
  13850=Cichlidae genus indeterminate

Family Embiotocidae (Surfperches)
  13900=Embiotocidae genus indeterminate

Family Eleotrididae (Sleepers)
  13950=Eleotrididae genus indeterminate

Family Gobiidae (Gobies)
  14000=Gobiidae genus indeterminate

Family Mugilidae (Mullets)
  14050=Mugilidae genus indeterminate

Family Percichthyidae (Temperate basses)
  14100=Percichthyidae genus indeterminate

Family Grammistidae (Soapfish)
  14150=Grammistidae genus indeterminate

Family Priacanthidae (Bigeyes)
  14200=Priacanthidae genus indeterminate

Family Apogonidae (Cardinalfish)
  14250=Apogonidae genus indeterminate

Family Malacanthidae (Tilefish)
  14300=Malacanthidae genus indeterminate

Family Pomatomidae (Bluefish)
  14350=Pomatomidae genus indeterminate

Family Rachycentridae (Cobia)
  14400=Rachycentridae genus indeterminate

Family Echeneididae (Sharksuckers)
  14450=Echeneididae genus indeterminate

Family Nematistiidae (Roosterfish)
  14500=Nematistiidae genus indeterminate

Family Coryphaenidae (Dolphin fish)
  14550=Coryphaenidae genus indeterminate

Family Bramidae (Pomfrets)
  14600=Bramidae genus indeterminate

Family Lutjanidae (Snappers)
  14650=Lutjanidae genus indeterminate

Family Lobotidae (Tripletails)
  14700=Lobotidae genus indeterminate

Family Gerreidae (Mojarras)
  14750=Gerreidae genus indeterminate

Family Haemulidae (Grunts)
  14800=Haemulidae genus indeterminate

Family Mullidae (Goatfish)
  14850=Mullidae genus indeterminate

Family Kyphosidae (Chubs or rudderfish)
  14900=Kyphosidae genus indeterminate

Family Ephippididae (Spadefish)
  14950=Ephippididae genus indeterminate

Family Chaetodontidae (Butterflyfish and Angelfish)
  15000=Chaetodontidae genus indeterminate

Family Pomacanthidae (Angelfish)
  15050=Pomacanthidae genus indeterminate

Family Pomacentridae (Damselfish)
  15100=Pomacentridae genus indeterminate

Family Cirrhitidae (Hawkfish)
  15150=Cirrhitidae genus indeterminate

Family Labridae (Wrasses)
  15200=Labridae genus indeterminate

Family Scaridae (Parrotfish)
  15250=Scaridae genus indeterminate

Family Stichaeidae (Pricklebacks)
  15300=Stichaeidae genus indeterminate

Family Cryptacanthodidae (Wrymouths)
  15350=Cryptacanthodidae genus indeterminate

Family Anarhichadidae (Wolffish)
  15400=Anarhichadidae genus indeterminate

Family Sphyraenidae (Barracudas)
  15450=Sphyraenidae genus indeterminate

Family Polynemidae (Threadfins)
  15500=Polynemidae genus indeterminate

Family Opistognathidae (Jawfish)
  15550=Opistognathidae genus indeterminate

Family Dactyloscopidae (Sand stargazers)
  15600=Dactyloscopidae genus indeterminate

Family Uranoscopidae (Stargazers)
  15650=Uranoscopidae genus indeterminate

Family Clinidae (Clinids)
  15700=Clinidae genus indeterminate

Family Blenniidae (Blennies)
  15750=Blenniidae genus indeterminate

Family Ammodytidae (Sand lances)
  15800=Ammodytidae genus indeterminate

Family Microdesmidae (Wormfish)
  15850=Microdesmidae genus indeterminate

Family Acanthuridae (Surgeonfish or tangs)
  15900=Acanthuridae genus indeterminate

Family Trichiuridae (Cutlassfish)
  15950=Trichiuridae genus indeterminate

Family Scombridae (Mackerels and tuna)
  16000=Scombridae genus indeterminate

Family Xiphiidae (Swordfish)
  16050=Xiphiidae genus indeterminate

Family Istiophoridae (Billfish)
  16100=Istiophoridae genus indeterminate

Family Stromateidae (Butterfish)
  16150=Stromateidae genus indeterminate

Family Ariommatidae (Ariommatids)
  16200=Ariommatidae genus indeterminate

Family Centrolophidae (Ruffs)
  16250=Centrolophidae genus indeterminate

Family Nomeidae (Driftfish)
  16300=Nomeidae genus indeterminate

Order Pleuronectiformes (Flounders, soles, etc.)
  16350=Pleuronectiformes family indeterminate

Family Bothidae (Flounders, dabs, etc.)
  16400=Bothidae genus indeterminate

Family Soleidae (Soles)
  16450=Soleidae genus indeterminate

Family Cynoglossidae (Tonguefish)
  16500=Cynoglossidae genus indeterminate

Family Pleuronectidae (Righteye flounders)
   16550=Pleuronectidae genus indeterminate

Order Tetraodontiformes (Triggerfish and allies)
   16600=Tetraodontiformes family indeterminate

   Family Balistidae (Triggerfish and filefish)
      16650=Balistidae genus indeterminate

   Family Ostraciidae (Trunkfish)
      16700=Ostraciidae genus indeterminate

   Family Tetraodontidae (Puffers)
      16750=Tetraodontidae genus indeterminate

   Family Diodontidae (Porcupine and burrfish)
      16800=Diodontidae genus indeterminate

   Family Molidae (Sunfish or headfish)
      16850=Molidae genus indeterminate

Order Scorpaeniformes
   Family Cottidae (Sculpins)
      16900=Cottidae genus indeterminate

Order Gasterosteiformes
   Family Gasterosteidae (Sticklebacks)
      17950=Gasterosteidae genus indeterminate

Class **Amphibia**
    30000=Amphibia order indeterminate

 Order Caudata (Salamanders and newts)
    30050=Caudata family indeterminate

    Family Cryptobranchidae
      30100=*Cryptobranchus* sp. (Hellbenders)

    Family Sirenidae
      30150=Sirenidae genus indeterminate
      30200=*Siren* sp. (Sirens)
      30250=*Pseudobranchus* sp. (Dwarf sirens)

    Family Ambystomatidae
      30300=*Ambystoma* sp. (Salamanders)

    Family Dicamptodontidae
      30350=Dicamptodontidae genus indeterminate
      30400=*Dicamptodon* sp. (Giant salamanders)
      30450=*Rhyacotriton* sp. (Olympic salamanders)

    Family Amphiumidae
      30500=*Amphiuma* sp. (Amphiumas)

    Family Plethodontidae
      30550=Plethodontidae genus indeterminate
      30600=*Aneides* sp. (Climbing salamanders)
      30650=*Batrachoseps* sp. (Slender salamanders)
      30700=*Desmognathus* sp. (Salamanders)
      30750=*Ensatina* sp. (Ensatinas)
      30800=*Eurycea* sp. (Salamanders)
      30850=*Gyrinophilus* sp. (Spring salamanders)
      30900=*Haideotriton* sp. (Blind salamanders)
      30950=*Hemidactylium* sp. (Four-toed salamanders)
      31000=*Hydromantes* sp. (Web-toed salamanders)
      31050=*Leurognathus* sp. (Shovel-nosed salamanders)
      31100=*Phaeognathus* sp. (Red hills salamander)
      31150=*Plethodon* sp. (Salamanders)
      31200=*Pseudotriton* sp. (Mud and red salamanders)
      31250=*Stereochilus* sp. (Many-lined salamanders)
      31300=*Typhlomolge* sp. (Salamanders)
      31350=*Typhlotriton* sp. (Grotto salamanders)

    Family Proteidae
      31400=*Necturus* sp. (Water dogs)

    Family Salamandridae
      31450=Salamandridae genus indeterminate
      31500=*Notophthalmus* sp. (Newts)
      31550=*Taricha* sp. (Newts)

 Order Anura (Frogs and toads)
    31600=Anura family indeterminate

    Family Rhinophrynidae
      31650=*Rhinophrynus* sp. (Burrowing toads)

Family Pelobatidae
  31700=*Scaphiopus* sp. (Spadefoot toads)

Family Ascaphidae
  31750=*Ascaphus* sp. (Tailed frogs)

Family Leptodactylidae
  31800=Leptodactylidae genus indeterminate
  31850=*Eleutherodactylus* sp. (Greenhouse frogs)
  31900=*Hylactophryne* sp. (Frogs)
  31950=*Leptodactylus* sp. (Frogs)
  32000=*Syrrhophus* sp. (Frogs)

Family Hylidae
  32050=Hylidae genus indeterminate
  32100=*Acris* sp. (Frogs)
  32150=*Hyla* sp. (Tree frogs)
  32200=*Limnaoedus* sp. (Grass frogs)
  32250=*Pseudacris* sp. (Chorus frogs)
  32300=*Pternohyla* sp. (Burrowing treefrogs)
  32350=*Smilisca* sp. (Tree frogs)

Family Bufonidae
  32400=*Bufo* sp. (Toads)

Family Ranidae
  32450=*Rana* sp. (Frogs)

Family Microhylidae
  32500=Microhylidae genus indeterminate
  32550=*Gastrophryne* sp. (Toads)
  32600=*Hypopachus* sp. (Frogs)

Family Pipidae
  32650=*Xenopus* sp. (Clawed frog)

Class **Reptilia**
    40010=Reptilia order indeterminate

Order Crocodilia (Crocodiles and alligators)
  Family Crocodilidae
    40100=*Alligator mississippiensis* (Alligator)
    40150=*Crocodylus acutus* (American crocodile)

Order Testudinata (Turtles and tortoises)
    40200=Testudinata family indeterminate (Turtles)

  Family Chelydridae (Snapping turtles)
    40250=Chelydridae genus indeterminate
    40300=*Chelydra serpentina* (Snapping turtle)
    40350=*Macroclemys temmincki* (Alligator snapping turtle)

  Family Kinosternidae (Mud and musk turtles)
    40400=Kinosternidae genus indeterminate
    40450=*Kinosternon* sp.
    40500=*Kinosternon bauri* (Striped mud turtle)
    40550=*Kinosternon flavescens* (Yellow mud turtle)
    40600=*Kinosternon hirtipes* (Mexican mud turtle)
    40650=*Kinosternon sonoriense* (Sonoran mud turtle)
    40700=*Kinosternon subrubrum* (Mud turtle)
    40750=*Sternotherus* sp.
    40800=*Sternotherus carinatus* (Razorback musk turtle)
    40850=*Sternotherus depressus* (Flattened musk turtle)
    40900=*Sternotherus minor* (Musk turtle)
    40950=*Sternotherus odoratus* (Stinkpot)

  Family Emydidae (Water and box turtles)
    41000=Emydidae genus indeterminate
    41050=*Chrysemys* sensu lato (*Chrysemys, Pseudemys,* & *Trachemys*)
    41100=*Chrysemys picta* (Painted turtle)
    41150=*Clemmys* sp.
    41200=*Clemmys guttata* (Spotted turtle)
    41250=*Clemmys insculpta* (Wood turtle)
    41300=*Clemmys marmorata* (Western pond turtle)
    41350=*Clemmys muhlenbergi* (Bog turtle)
    41400=*Deirochelys reticularia* (Chicken turtle)
    41450=*Emydoidea blandingi* (Blanding's turtle)
    41500=*Graptemys* sp.
    41550=*Graptemys barbouri* (Barbour's map turtle)
    41600=*Graptemys caglei* (Cagle's map turtle)
    41650=*Graptemys flavimaculata* (Yellow-blotched sawback)
    41700=*Graptemys geographica* (Map turtle)
    41750=*Graptemys kohni* (Mississippi map turtle)
    41800=*Graptemys nigrinoda* (Black-knobbed sawback)
    41850=*Graptemys oculifera* (Ringed sawback)
    41900=*Graptemys pseudogeographica* (False map turtle)
    41950=*Graptemys pulchra* (Alabama map turtle)
    42000=*Graptemys versa* (Texas map turtle)
    42050=*Malaclemys terrapin* (Diamondback terrapin)
    42100=*Pseudemys* sp.

            42150=*Pseudemys concinna* (River cooter)
            42200=*Pseudemys floridana* (Florida cooter)
            42250=*Pseudemys nelsoni* (Florida red-bellied turtle)
            42300=*Pseudemys texana* (Texas river cooter)
            42350=*Terrapene* sp.
            42400=*Terrapene carolina* (Box turtle)
            42450=*Terrapene ornata* (Box turtle)
            42500=*Trachemys* sp.
            42550=*Trachemys gaigeae* (Big Bend slider)
            42600=*Trachemys scripta* (Slider)

        Family Testudinidae
            42650=*Gopherus* sp.
            42700=*Gopherus agassizii* (Desert tortoise)
            42750=*Gopherus berlandieri* (Texas tortoise)
            42780=*Gopherus polyphemus* (Gopher tortoise)

        Family Trionychidae
            42800=*Trionyx* sp.
            42850=*Trionyx muticus* (Smooth softshell)
            42900=*Trionyx spiniferus* (Spiny softshell)

        Family Chelonidae (Marine turtles)
            42950=Chelonidae genus indeterminate
            43000=*Caretta caretta* (Loggerhead)
            43050=*Chelonia mydas* (Green turtle)
            43100=*Eretmochelys imbricata* (Hawksbill)
            43150=*Lepidochelys* sp.
            43200=*Lepidochelys kempi* (Atlantic ridley)
            43250=*Lepidochelys olivacea* (Pacific ridley)

        Family Dermochelyidae
            43300=*Dermochelys coriacea* (Leatherback)

Order Squamata
            43350=Squamata suborder indeterminate (Lizards and
                snakes)

    Suborder Lacertilia
            43400=Lacertilia family indeterminate (Lizards)

        Family Gekkonidae
            43450=Gekkonidae genus indeterminate
            43500=*Coleonyx* sp. (Gekkos)
            43550=*Cyrtodactylus* sp. (Rough-scaled gekkos)
            43600=*Gonatodes* sp. (Yellow-headed gekkos)
            43650=*Hemidactylus* sp. (Mediterranean gekkos)
            43700=*Phyllodactylus* sp. (Leaf-toed gekkos)
            43750=*Sphaerodactylus* sp. (Gekkos)

        Family Iguanidae
            43800=Iguanidae genus indeterminate
            43850=*Anolis* sp. (Anoles)
            43900=*Callisaurus* sp. (Lizards)
            43950=*Cophosaurus* sp. (Greater earless lizards)
            44000=*Crotaphytus* sp. (Collared lizards)
            44050=*Ctenosaura* sp. (Spiny-tailed iguanas)

        44100=*Dipsosaurus* sp. (Desert iguanas)
        44150=*Gambelia* sp. (Leopard lizards)
        44200=*Holbrookia* sp. (Earless lizards)
        44250=*Leiocephalus* sp. (Curly-tailed lizards)
        44300=*Petrosaurus* sp. (Rock lizards)
        44350=*Phrynosoma* sp. (Horned lizards)
        44400=*Sauromalus* sp. (Chuckwallas)
        44450=*Sceloporus* sp. (Spiny lizards)
        44500=*Uma* sp. (Fringe-toed lizard)
        44550=*Urosaurus* sp. (Tree lizards)
        44600=*Uta* sp. (Side-blotched lizards)

      Family Xantusiidae
        44650=*Xantusia* sp. (Night lizards)

      Family Scincidae
        44700=Scincide genus indeterminate
        44750=*Eumeces* sp. (Skinks)
        44800=*Leiolopisma* sp. (Skinks)
        44850=*Neoseps* sp. (Sand skinks)
        44900=*Scincella* sp. (Ground skinks)

      Family Teiidae
        44950=*Cnemidophorus* sp. (Whiptail lizards)

      Family Anguinidae
        45000=Anguinidae genus indeterminate
        45050=*Gerrhonotus* sp. (Alligator lizards)
        45100=*Ophisaurus* sp. (Glass lizards)

      Family Anniellidae
        45150=*Anniella* sp. (Legless lizards)

      Family Helodermatidae
        45200=*Heloderma* sp. (Gila monster)

    Suborder Amphisbaenia
      Family Amphisbaenidae
        45250=*Rhineura* sp. (Worm lizards)

    Suborder Serpentes
        45300=Serpentes family indeterminate (Snakes)

      Family Leptotyphlopidae
        45350=*Leptotyphlops* sp. (Blind snakes)

      Family Boidae
        45400=Boidae genus indeterminate
        45450=*Charina* sp. (Boa)
        45500=*Lichanura* sp. (Boa)

      Family Colubridae
        45550=Colubridae genus indeterminate
        45600=*Arizona* sp. (Glossy snakes)
        45650=*Carphophis* sp. (Worm snakes)
        45700=*Cemophora* sp. (Scarlet snakes)
        45750=*Chilomeniscus* sp. (Sand snakes)

45800=*Chionactis* sp. (Shovel-nosed snakes)
45850=*Coluber* sp. (Racers)
45900=*Coniophanes* sp. (Black-striped snakes)
45950=*Contia* sp. (Sharp-tailed snakes)
46000=*Diadophis* sp. (Ringneck snakes)
46050=*Drymarchon* sp. (Indigo snakes)
46100=*Drymobius* sp. (Speckled racers)
46150=*Elaphe* sp. (Corn and rat snakes)
46200=*Eridiphas* sp. (Night snakes)
46250=*Farancia* sp. (Mud and rainbow snakes)
46300=*Ficimia* sp. (Mexican hooknose snakes)
46350=*Gyalopion* sp. (Western hooknose snakes)
46400=*Heterodon* sp. (Hognose snakes)
46450=*Hypsiglena* sp. (Night snakes)
46500=*Lampropeltis* sp. (Kingsnakes)
46550=*Leptodeira* sp. (Cat-eyed snakes)
46600=*Liodytes* sp. (Swamp snakes)
46650=*Masticophis* sp. (Whip snakes)
46700=*Nerodia* sp. (Water snakes)
46750=*Opheodrys* sp. (Green snakes)
46800=*Oxybelis* sp. (Vine snakes)
46850=*Phyllorhynchus* sp. (Leaf-nosed snakes)
46900=*Pituophis* sp. (Bullsnakes)
46950=*Rhadinaea* sp. (Pine woods snakes)
47000=*Regina* sp. (Crayfish snakes)
47050=*Rhinocheilus* sp. (Longnose snakes)
47100=*Salvadora* sp. (Patchnose snakes)
47150=*Seminatrix* sp. (Swamp snakes)
47200=*Sonora* sp. (Ground snakes)
47250=*Stilosoma* sp. (Short-tailed snakes)
47300=*Storeria* sp. (Brown and redbelly snakes)
47350=*Tantilla* sp. (Blackhead snakes)
47400=*Thamnophis* sp. (Garter snakes)
47450=*Trimorphodon* sp. (Lyre snakes)
47500=*Tropidoclonion* sp. (Lined snakes)
47550=*Virginia* sp. (Earth snakes)

Family Elapidae
47600=*Micrurus* sp. (Coral snakes)

Family Hydrophiidae
47650=*Pelamis* sp. (Sea snakes)

Family Viperidae
47700=Viperidae genus indeterminate
47750=*Agkistrodon* sp. (Copperhead and cottonmouth snakes)
47800=*Crotalus* sp. (Rattlesnakes)
47850=*Sistrurus* sp. (Massasauga and pygmy rattlesnake)

Class **Aves** (Birds)
    50010=Small bird indeterminate (Small perching birds or smaller)
    50020=Small/medium bird indeterminate
    50030=Medium bird indeterminate (Large perching birds, etc.)
    50040=Medium/large bird indeterminate
    50050=Large bird indeterminate (Vultures, turkeys, ducks, etc.)
    50060=Aves, size indeterminate

Order **Gaviiformes** (Loons)
  Family Gaviidae
    50100=*Gavia* sp. (Loons)

Order **Podicipediformes** (Grebes)
  Family Podicipedidae
    50150=Podicipedidae genus indeterminate
    50200=*Aechmophorus* sp.
    50250=*Podiceps* sp.
    50300=*Podilymbus* sp.
    50350=*Tachybaptus* sp.

Order **Procellariiformes** (Fulmers, shearwaters, petrels)
    50400=Procellariiformes family indeterminate

  Family Diomedeidae
    50450=*Diomedea* sp. (Albatrosses)

  Family Procellariidae
    50500=Procellariidae genus indeterminate
    50550=*Calonectris* sp. (Shearwaters)
    50600=*Fulmarus* sp. (Fulmers)
    50650=*Pterodroma* sp. (Petrels)
    50700=*Puffinus* sp. (Shearwaters)

  Family Hydrobatidae
    50750=*Oceanodroma* sp. (Storm-petrels)

Order **Pelicaniformes** (Pelicans and allies)
    50800=Pelicaniformes family indeterminate

  Family Phaethontidae
    50850=*Phaethon* sp. (Tropicbirds)

  Family Pelecanidae
    50900=*Pelecanus* sp. (Pelicans)

  Family Fregatidae
    50950=*Fregata* sp. (Frigate birds)

  Family Sulidae
    51000=*Sula* sp. (Gannets and Boobies)

  Family Phalacrocoracidae (Cormorants)
    51050=*Phalacrocorax* sp.

Family Anhingidae (Anhingas)
  51100=*Anhinga* sp.

Order Ciconiiformes (Herons, egrets and bitterns)
  51150=Ciconiiformes family indeterminate

  Family Ardeidae
    51200=Ardeidae genus indeterminate
    51250=*Ardea* sp. (Herons)
    51300=*Botaurus* sp. (Bitterns)
    51350=*Bubulcus* sp. (Egrets)
    51400=*Butorides* sp. (Herons)
    51450=*Casmerodius* sp. (Egrets)
    51500=*Egretta* sp. (Egrets and herons)
    51550=*Ixobrychus* sp. (Bitterns)
    51560=*Nyctanassa* sp. (Herons)
    51600=*Nycticorax* sp. (Herons)

  Family Threskiornithidae
    51650=Threskiornithidae genus indeterminate
    51700=*Ajaia* sp. (Spoonbills)
    51750=*Eudocimus* sp. (Ibis)
    51800=*Plegadis* sp. (Ibis)

  Family Ciconiidae
    51850=*Mycteria* sp. (Storks)

  Family Phoenicopteridae
    51900=*Phoenicopterus* sp. (Flamingoes)

Order Anseriformes (Swans, geese, and ducks)
  Family Anatidae
    51950=Anatidae genus indeterminate
    52000=*Aix* sp. (Ducks)
    52050=*Anas* sp. (Ducks)
    52100=*Anser* sp. (Geese)
    52150=*Aythya* sp. (Ducks)
    52200=*Branta* sp. (Brants)
    52250=*Bucephala* sp. (Ducks)
    52300=*Chen* sp. (Geese)
    52350=*Clangula* sp. (Ducks)
    52400=*Cygnus* sp. (Swans)
    52450=*Dendrocygna* sp. (Whistling-ducks)
    52500=*Histrionicus* sp. (Ducks)
    52550=*Lophodytes* sp. (Ducks)
    52600=*Melanitta* sp. (Ducks)
    52650=*Mergellus* sp. (Smews)
    52700=*Mergus* sp. (Ducks)
    52750=*Oxyura* sp. (Ducks)
    52800=*Somateria* sp. (Eiders)

Order Falconiformes (Vultures, hawks, and falcons)
  52850=Falconiformes family indeterminate

  Family Cathartidae
    52900=Cathartidae genus indeterminate
    52950=*Cathartes* sp. (Vultures)

```
        53000=Coragyps sp. (Vultures)
        53050=Gymnogyps sp. (Condors)

    Family Accipitridae
        53100=Accipitridae genus indeterminate
        53150=Accipiter sp. (Hawks)
        53200=Aquila sp. (Eagles)
        53250=Buteo sp. (Hawks)
        53300=Buteogallus sp. (Hawks)
        53350=Chondrohierax sp. (Kites)
        53400=Circus sp. (Harriers)
        53450=Elanoides sp. (Kites)
        53500=Elanus sp. (Kites)
        53550=Haliaeetus sp. (Eagles)
        53600=Ictinia sp. (Kites)
        53650=Pandion sp. (Ospreys)
        53700=Parabuteo sp. (Hawks)
        53750=Rostrhamus sp. (Kites)

    Family Falconidae
        53800=Falconidae genus indeterminate
        53850=Falco sp. (Falcons)
        53900=Polyborus sp. (Caracaras)

Order Galliformes (Turkeys, grouse, quails, and
            chachalacas)
        53950=Galliformes family indeterminate

    Family Cracidae
        54000=Ortalis sp. (Chachalacas)

    Family Phasianidae
        54050=Phasianidae genus indeterminate
        54100=Alectoris sp. (Chukars)
        54150=Bonasa sp. (Grouse)
        54200=Callipepla sp. (Quail)
        54250=Centrocerus sp. (Grouse)
        54300=Colinus sp. (Bobwhites)
        54350=Cyrtonyx sp. (Quail)
        54400=Dendragapus sp. (Grouse)
        54410=Gentriops sp. (Southern coastal "turkey")
        54450=Francolinus sp. (Francolins)
        54500=Gallus gallus (Domestic chicken)
        54550=Lagopus sp. (Ptarmigans)
        54600=Meleagris gallopavo (Turkeys)
        54650=Oreotyx sp. (Quail)
        54700=Perdix sp. (Partridges)
        54750=Phasianus sp. (Pheasants)
        54800=Tympanuchus sp. (Prairie-chickens and grouse)

    Family Numididae
        54850=Numida sp. (Guineafowl)
```

Order Gruiformes (Cranes and allies)
  54900=Gruiformes family indeterminate

  Family Rallidae
    54950=Rallidae genus indeterminate
    55000=*Coturnicops* sp. (Rails)
    55050=*Crex* sp. (Crakes)
    55100=*Fulica* sp. (Coots)
    55150=*Gallinula* sp. (Moorhens)
    55200=*Laterallus* sp. (Rails)
    55250=*Porphyrula* sp. (Gallinules)
    55300=*Porzana* sp. (Soras)
    55350=*Rallus* sp. (Rails)

  Family Aramidae
    55400=*Aramus* sp. (Limpkins)

  Family Gruidae
    55450=*Grus* sp. (Cranes)

Order Charadriiformes (Shore birds, gulls, and alcids)
  55500=Charadriiformes family indeterminate

  Family Charadriidae
    55550=Charadriidae genus indeterminate
    55600=*Charadrius* sp. (Plovers)
    55650=*Pluvialis* sp. (Plovers)

  Family Haematopodidae
    55700=*Haematopus* sp. (Oystercatchers)

  Family Recurvirostridae
    55750=Recurvirostridae genus indeterminate
    55800=*Himantopus* sp. (Stilts)
    55850=*Recurvirostra* sp. (Avocets)

  Family Jacanidae
    55900=*Jacana* sp. (Jacanas)

  Family Scolopacidae
    55950=Scolopacidae genus indeterminate
    56000=*Actitis* sp. (Sandpipers)
    56050=*Aphriza* ap. (Surfbirds)
    56100=*Arenaria* sp. (Turnstones)
    56150=*Bartramia* sp. (Sandpipers)
    56200=*Calidris* sp. (Knots, sanderlings, sandpipers, and dunlins)
    56250=*Catoptrophorus* sp. (Willets)
    56300=*Eurynorhynchus* sp. (Sandpipers)
    56350=*Gallinago* sp. (Snipes)
    56400=*Heteroscelus* sp. (Tattlers)
    56450=*Limicola* sp. (Sandpipers)
    56500=*Limnodromus* sp. (Dowitchers)
    56550=*Limosa* sp. (Godwits)
    56600=*Numenius* sp. (Curlews and whimbrels)
    56650=*Phalaropus* sp. (Phalaropes)
    56700=*Scolopax* sp. (Woodcocks)

    56750=*Tringa* sp. (Yellowlegs and sandpipers)
    56800=*Tryngites* sp. (Sandpipers)
    56850=*Xenus* sp. (Sandpipers)

  Family Laridae
    56900=Laridae genus indeterminate
    56950=*Anous* sp. (Noddies)
    57000=*Catharacta* sp. (Skuas)
    57050=*Chlidonias* sp. (Terns)
    57100=*Larus* sp. (Gulls)
    57150=*Pagophila* sp. (Gulls)
    57200=*Rhodostethia* sp. (Gulls)
    57250=*Rissa* sp. (Kittiwakes)
    57300=*Rynchops* sp. (Skimmers)
    57350=*Stercorarius* sp. (Jaegers)
    57400=*Sterna* sp. (Terns)
    57450=*Xema* sp. (Gulls)

  Family Alcidae
    57500=Alcidae genus indeterminate
    57550=*Aethia* sp. (Auklets)
    57600=*Alca* sp. (Razorbills)
    57650=*Alle* sp. (Dovekies)
    57700=*Brachyramphus* sp. (Murrelets)
    57750=*Cepphus* sp. (Guillemots)
    57800=*Cerorhinca* sp. (Auklets)
    57850=*Cyclorrhynchus* sp. (Auklets)
    57900=*Fratercula* sp. (Puffins)
    57950=*Ptychoramphus* sp. (Auklets)
    58000=*Synthliboramphus* sp. (Murrelets)
    58050=*Uria* sp. (Murres)

Order Columbiformes (Pigeons and doves)
  Family Columbidae
    58100=Columbidae genus indeterminate
    58150=*Columba* sp. (Doves and pigeons)
    58200=*Columbina* sp. (Doves)
    58250=*Ectopistes* sp. (Pigeons)
    58300=*Leptotila* sp. (Doves)
    58350=*Streptopelia* sp. (Doves)
    58400=*Zenaida* sp. (Doves)

Order Psittaciformes (Parrots and parakeets)
  Family Psittacidae
    58450=Psittacidae genus indeterminate
    58500=*Amazona* sp. (Parrots)
    58530=*Ara* sp. (Macaws)
    58550=*Aratinga* sp. (Parakeets)
    58600=*Brotogeris* sp. (Parakeets)
    58650=*Conuropsis* sp. (Parakeets)
    58700=*Melopsittacus* sp. (Budgerigars)
    58750=*Myiopsitta* sp. (Parakeets)
    58800=*Nandayus* sp. (Conures)
    58850=*Psittacula* sp. (Parakeets)

Order Cuculiformes (Cuckoos, anis, and roadrunners)
  Family Cuculidae
    58900=Cuclidae genus indeterminate
    58950=*Coccyzus* sp. (Cuckoos)
    59000=*Crotophaga* sp. (Ani)
    59050=*Geococcyx* sp. (Roadrunners)

Order Strigiformes (Owls)
    59100=Strigiformes family indeterminate

  Family Tytonidae
    59150=*Tyto* sp. (Owls)

  Family Strigidae
    59200=Strigidae genus indeterminate
    59250=*Aegolius* sp. (Owls)
    59300=*Asio* sp. (Owls)
    59350=*Athene* sp. (Owls)
    59400=*Bubo* sp. (Owls)
    59450=*Glaucidium* sp. (Owls)
    59500=*Micrathene* sp. (Owls)
    59550=*Nyctea* sp. (Owls)
    59600=*Otus* sp. (Owls)
    59650=*Surnia* sp. (Owls)
    59700=*Strix* sp. (Owls)

Order Caprimulgiformes (Goatsuckers)
  Family Caprimulgidae
    59750=Caprimulgidae genus indeterminate
    59800=*Caprimulgus* sp. (Whip-poor-wills and
        Chuck-wills'-widow)
    59850=*Chordeiles* sp. (Nighthawks)
    59900=*Nyctidromus* sp. (Pauraques)
    59950=*Phalaenoptilus* sp. (Poorwills)

Order Apodiformes (Swifts and hummingbirds)
    60000=Apodiformes family indeterminate

  Family Apodidae
    60050=Apodidae genus indeterminate
    60100=*Aeronautes* sp. (Swifts)
    60150=*Chaetura* sp. (Swifts)
    60200=*Cypseloides* sp. (Swifts)

  Family Trochilidae
    60250=Trochilidae genus indeterminate
    60300=*Amazilia* sp. (Hummingbirds)
    60350=*Archilochus* sp. (Hummingbirds)
    60400=*Calothorax* sp. (Hummingbirds)
    60450=*Calypte* sp. (Hummingbirds)
    60500=*Chlorostilbon* sp. (Hummingbirds)
    60550=*Cynanthus* sp. (Hummingbirds)
    60600=*Eugenes* sp. (Hummingbirds)
    60650=*Hylocharis* sp. (Hummingbirds)
    60700=*Lampornis* sp. (Hummingbirds)
    60750=*Selasphorus* sp. (Hummingbirds)
    60800=*Stellula* sp. (Hummingbirds)

Order Trogoniformes (Trogons)
  Family Trogonidae
    60850=*Trogon* sp. (Trogons)

Order Coraciiformes (Kingfishers)
  Family Alcedinidae
    60900=Alcedinidae genus indeterminate
    60950=*Ceryle* sp. (Kingfishers)
    61000=*Chloroceryle* sp. (Kingfishers)

Order Piciformes (Woodpeckers)
  Family Picidae
    61050=Picidae genus indeterminate
    61100=*Campephilus* sp. (Woodpeckers)
    61150=*Colaptes* sp. (Woodpeckers)
    61200=*Dryocopus* sp. (Woodpeckers)
    61250=*Melanerpes* sp. (Woodpeckers)
    61300=*Picoides* sp. (Woodpeckers)
    61350=*Sphyrapicus* sp. (Woodpeckers)

Order Passeriformes (Perching birds)
    61400=Passeriformes family indeterminate

  Family Tyrannidae
    61450=Tyrannidae genus indeterminate
    61500=*Camptostoma* sp. (Tyrannulets)
    61550=*Contopus* sp. (Flycatchers and pewees)
    61600=*Empidonax* sp. (Flycatchers)
    61650=*Myiarchus* sp. (Flycatchers)
    61700=*Myiodynastes* sp. (Flycatchers)
    61750=*Pachyramphus* sp. (Becards)
    61800=*Pitangus* sp. (Kiskadees)
    61850=*Pyrocephalus* sp. (Flycatchers)
    61900=*Sayornis* sp. (Phoebes)
    61950=*Tyrannus* sp. (Kingbirds)

  Family Alaudidae
    62000=Alaudidae genus indeterminate
    62050=*Alauda* sp. (Skylarks)
    62100=*Eremophila* sp. (Larks)

  Family Hirundinidae
    62150=Hirundinidae genus indeterminate
    62200=*Hirundo* sp. (Swallows)
    62250=*Progne* sp. (Martins)
    62300=*Riparia* sp. (Swallows)
    62350=*Stelgidopteryx* sp. (Swallows)
    62400=*Tachycineta* sp. (Swallows)

  Family Corvidae
    62450=Corvidae genus indeterminate
    62500=*Aphelocoma* sp. (Jays)
    62550=*Corvus* sp. (Crows and ravens)
    62600=*Cyanocitta* sp. (Jays)
    62650=*Cyanocorax* sp. (Jays)
    62700=*Gymnorhinus* sp. (Jays)
    62750=*Nucifraga* sp. (Nutcrackers)

      62800=*Perisoreus* sp. (Jays)
      62850=*Pica* sp. (Magpies)

    Family Paridae
      62900=*Parus* sp. (Chickadees and titmice)

    Family Remizidae
      62950=*Auriparus* sp. (Verdins)

    Family Aegithalidae
      63000=*Psaltriparus* sp. (Bushtits)

    Family Pycnonotidae
      63050=*Pycnonotus* sp. (Bulbuls)

    Family Sittidae
      63100=*Sitta* sp. (Nuthatches)

    Family Cinclidae
      63150=*Cinclus* sp. (Dippers)

    Family Certhiidae
      63200=*Certhia* sp. (Creepers)

    Family Troglodytidae
      63250=Troglodytidae genus indeterminate
      63300=*Campylorhynchus* sp. (Wrens)
      63350=*Catherpes* sp. (Wrens)
      63400=*Cistothorus* sp. (Wrens)
      63450=*Salpinctes* sp. (Wrens)
      63500=*Thryomanes* sp. (Wrens)
      63550=*Thyothorus* sp. (Wrens)
      63600=*Troglodytes* sp. (Wrens)

    Family Muscicapidae
      63650=Muscicapidae genus indeterminate
      63700=*Catharus* sp. (Veeries and thrushes)
      63750=*Chamaea* sp. (Wrentits)
      63800=*Hylocichla* sp. (Thrushes)
      63850=*Ixoreus* sp. (Thrushes)
      63900=*Luscinia* sp. (Rubythroats)
      63950=*Muscicapa* sp. (Flycatchers)
      64000=*Myadestes* sp. (Solitaires)
      64050=*Oenanthe* sp. (Wheatwears)
      64100=*Phylloscopus* sp. (Warblers)
      64150=*Polioptila* sp. (Gnatcatchers)
      64200=*Regulus* sp. (Kinglets)
      64250=*Sialia* sp. (Bluebirds)
      64300=*Turdus* sp. (Robins)

    Family Mimidae
      64350=Mimidae genus indeterminate
      64400=*Dumetella* sp. (Catbirds)
      64450=*Mimus* sp. (Mockingbirds)
      64500=*Oreoscoptes* sp. (Thrashers)
      64550=*Toxostoma* sp. (Thrashers)

Family Motacillidae
  64600=Motacillidae genus indeterminate
  64650=*Anthus* sp. (Pipits)
  64700=*Motacilla* sp. (Wagtails)

Family Bombycillidae
  64750=*Bombycilla* sp. (Waxwings)

Family Ptilogonatidae
  64800=*Phainopepla* sp. (Phainopeplas)

Family Laniidae
  64850=*Lanius* sp. (Shrikes)

Family Sturnidae
  64900=Sturnidae genus indeterminate
  64950=*Acridotheres* sp. (Mynas)
  65000=*Sturnus* sp. (Starlings)

Family Vireonidae
  65050=*Vireo* sp. (Vireos)

Family Emberizidae
  65100=Emberizidae genus indeterminate
  65150=*Agelaius* sp. (Blackbirds)
  65200=*Aimophila* sp. (Sparrows)
  65250=*Ammodramus* sp. (Sparrows)
  65300=*Amphispiza* sp. (Sparrows)
  65350=*Arremonops* sp. (Sparrows)
  65400=*Calamospiza* sp. (Buntings)
  65450=*Calcarius* sp. (Longspurs)
  65500=*Cardellina* sp. (Warblers)
  65550=*Cardinalis* sp. (Cardinals)
  65650=*Chondestes* sp. (Sparrows)
  65700=*Coereba* sp. (Bananaquits)
  65750=*Dendroica* sp. (Warblers)
  65800=*Dolichonyx* sp. (Bobolinks)
  65850=*Euphagus* sp. (Blackbirds)
  65900=*Geothlypis* sp. (Yellowthroats)
  65950=*Guiraca* sp. (Grosbeaks)
  66000=*Helmitheros* sp. (Warblers)
  66050=*Icterus* sp. (Orioles)
  66100=*Icteria* sp. (Chats)
  66150=*Junco* sp. (Juncos)
  66200=*Limnothlypis* sp. (Warblers)
  66250=*Melospiza* sp. (Sparrows)
  66300=*Mniotilta* sp. (Warblers)
  66350=*Molothrus* sp. (Cowbirds)
  66400=*Myioborus* sp. (Redstarts)
  66450=*Oporornis* sp. (Warblers)
  66500=*Parula* sp. (Warblers)
  66550=*Passerculus* sp. (Sparrows)
  66600=*Passerella* sp. (Sparrows)
  66650=*Passerina* sp. (Buntings)
  66700=*Peucedramus* sp. (Warblers)
  66750=*Pheucticus* sp. (Grosbeaks)
  66800=*Pipilo* sp. (Towhees)

66850=*Piranga* sp. (Tanagers)
66900=*Plectrophenax* sp. (Longspurs)
66950=*Pooecetes* sp. (Sparrows)
67000=*Protonotaria* sp. (Warblers)
67050=*Quiscalus* sp. (Grackles)
67100=*Seiurus* sp. (Ovenbirds and waterthrushes)
67150=*Setophaga* sp. (Redstarts)
67200=*Spindalis* sp. (Tanager)
67250=*Spiza* sp. (Dickcissels)
67300=*Spizella* sp. (Sparrows)
67350=*Sporophila* sp. (Seedeaters)
67400=*Sturnella* sp. (Meadowlarks)
67450=*Vermivora* sp. (Warblers)
67500=*Wilsonia* sp. (Warblers)
67550=*Xanthocephalus* sp. (Blackbirds)
67600=*Zonotrichia* sp. (Sparrows)

Family Fringillidae
67650=Fringillidae genus indeterminate
67700=*Coccothraustes* sp. (Grosbeaks)
67750=*Carduelis* sp. (Siskins and goldfinches)
67800=*Carpodacus* sp. (Finches)
67850=*Loxia* sp. (Crossbills)

Family Passeridae
67900=*Passer* sp. (Sparrows)

Class Mammalia

    70010=Micro mammal indeterminate (Species less than 100 grams)
    70020=Micro/small mammal indeterminate
    70030=Small mammal indeterminate (*S. floridanus* to 100 grams)
    70040=Small/medium mammal indeterminate
    70050=Medium mammal indeterminate (Canid, caprine-sized)
    70060=Medium/large mammal indeterminate
    70070=Large mammal indeterminate (Deer, pronghorn, bighorn-sized)
    70080=Large/very large mammal indeterminate
    70090=Very large mammal indeterminate (Elk, bison-sized or larger)
    70100=Mammalia, size indeterminate

Order Marsupialia (Opossums)
  Family Didelphidae
    70150=*Didelphis virginiana* (Virginia opossum)

Order Soricomorpha (Insectivora) (Insectivores)
    70200=Soricomorpha family indeterminate

  Family Soricidae
    70250=Soricidae genus indeterminate
    70300=*Blarina* sp.
    70350=*Blarina brevicauda* (Short-tailed shrew)
    70400=*Blarina carolinensis* (Southern short-tailed shrew)
    70430=*Blarina hylophaga* (Short-tailed shrew)
    70450=*Blarina telmalestes* (Swamp short-tailed shrew)
    70500=*Cryptotis parva* (Least shrew)
    70550=*Microsorex hoyi* (Pygmy shrew)
    70600=*Notiosorex crawfordi* (Desert shrew)
    70650=*Sorex* sp.
    70700=*Sorex alaskanus* (Glacier Bay water shrew)
    70750=*Sorex arcticus* (Saddle-backed or arctic shrew)
    70800=*Sorex arizonae* (Arizona shrew)
    70850=*Sorex bendirii* (Pacific water shrew)
    70900=*Sorex cinereus* (Masked or cinereus shrew)
    70950=*Sorex dispar* (Long-tailed or rock shrew)
    71000=*Sorex fumeus* (Smoky shrew)
    71050=*Sorex gaspensis* (Gaspe shrew)
    71100=*Sorex hydrodromus* (Unalaska shrew)
    71150=*Sorex juncensis* (Tule shrew)
    71200=*Sorex longirostris* (Bachman's shrew)
    71250=*Sorex lyelli* (Mount Lyell shrew)
    71300=*Sorex merriami* (Merriam's shrew)
    71350=*Sorex milleri* (Carmen Mountain shrew)
    71400=*Sorex monticola* (Dusky shrew)
    71450=*Sorex nanus* (Dwarf shrew)
    71500=*Sorex ornatus* (Ornate shrew)
    71550=*Sorex pacificus* (Pacific shrew)
    71600=*Sorex palustris* (Water shrew)
    71650=*Sorex preblei* (Malheur shrew)

71700=*Sorex pribilofensis* (Pribilof shrew)
71750=*Sorex sinuosus* (Suidun shrew)
71800=*Sorex tenellus* (Inyo shrew)
71850=*Sorex trigonirostris* (Ashland shrew)
71900=*Sorex trowbridgii* (Trowbridge's shrew)
71950=*Sorex vagrans* (Vagrant shrew)
72000=*Sorex willetti* (Santa Catalina shrew)

Family Talpidae
72050=Talpidae genus indeterminate
72100=*Condylura cristata* (Star-nosed mole)
72150=*Neurotrichus gibbsii* (Shrew-mole)
72200=*Parascalops breweri* (Hairy-tailed mole)
72250=*Scapanus* sp.
72300=*Scapanus latimanus* (Broad-footed mole)
72350=*Scapanus orarius* (Coast mole)
72400=*Scapanus townsendii* (Townsend's mole)
72450=*Scalopus aquaticus* (Eastern mole)

Order Chiroptera (Bats)
72500=Chiroptera family indeterminate

Family Mormoopidae
72550=*Mormoops megalophylla* (Leaf-chinned or ghost-faced bat)

Family Phyllostomatidae
72600=Phyllostomatidae genus indeterminate
72650=*Choeronycteris mexicana* (Long-tongued bat)
72700=*Diphylla ecaudata* (Hairy-legged vampire bat)
72750=*Leptonycteris* sp.
72800=*Leptonycteris nivalis* (Mexican long-tongued bat)
72850=*Leptonycteris sanborni* (Sanborn's long-nosed bat)
72900=*Macrotus californicus* (California leaf-nosed bat)

Family Vespertilionidae (Insectivorous bats)
72950=Vespertilionidae genus indeterminate
73000=*Myotis* sp.
73050=*Myotis auriculus* (Southwestern Myotis)
73100=*Myotis austroriparius* (Southeastern bat)
73150=*Myotis californicus* (California bat)
73200=*Myotis evotis* (Long-eared bat)
73250=*Myotis grisescens* (Gray Myotis)
73300=*Myotis keenii* (Keen's Myotis)
73350=*Myotis leibii* (Masked bat)
73400=*Myotis lucifugus* (Little brown bat)
73450=*Myotis occultus* (Arizona Myotis)
73500=*Myotis thysanodes* (Fringed bat)
73550=*Myotis sodalis* (Indiana Myotis or social Myotis)
73600=*Myotis subulatus* (Small-footed Myotis)
73650=*Myotis velifer* (Cave bat)
73700=*Myotis volans* (Long-legged bat)
73750=*Myotis yumanensis* (Yuma bat)

        73800=*Lasionycteris noctivagans* (Silver-haired bat)
        73850=*Lasiurus* sp.
        73900=*Lasiurus borealis* (Red bat)
        73950=*Lasiurus cinereus* (Hoary bat)
        74000=*Lasiurus ega* (Lesser yellow bat)
        74050=*Lasiurus humeralis* (Evening bat)
        74100=*Lasiurus intermedius* (Greater yellow bat)
        74150=*Lasiurus seminolus* (Seminole bat)
        74200=*Pipistrellus* sp.
        74250=*Pipistrellus hesperus* (Western canyon bat)
        74300=*Pipistrellus subflavus* (Georgia bat)
        74350=*Eptesicus fuscus* (Big brown bat)
        74400=*Euderma maculatum* (Spotted bat)
        74450=*Plecotus* sp.
        74500=*Plecotus phyllotis* (Mexican big-eared bat)
        74550=*Plecotus rafinesquii* (Rafinesque's big-eared
            bat)
        74600=*Plecotus townsendii* (Townsend's big-eared bat)
        74650=*Antrozous pallidus* (Pallid bat)
        74700=*Nycticeius humeralis* (Evening bat)
        74750=*Idionycteris phyllotis* (Allen's big-eared bat)

    Family Molossidae (Free-tailed bats)
        74800=Molossidae genus indeterminate
        74850=*Tadarida* sp.
        74900=*Tadarida brasiliensis* (Brazilian free-tailed
            bat)
        74950=*Tadarida cynocephala* (Florida free-tailed bat)
        75000=*Tadarida femorosacca* (Pocketed free-tailed
            bat)
        75050=*Tadarida macrotis* (Big free-tailed bat)
        75100=*Tadarida mexicana* (Guano bat)
        75150=*Eumops* sp.
        75200=*Eumops glaucinus* (Wagner's mastiff bat)
        75250=*Eumops perotis* (Western mastiff bat)
        75300=*Eumops underwoodi* (Underwood's mastiff bat)

Order Xenarthra (Edentata) (Armadillos)
    Family Dasypodidae
        75350=*Dasypus novemcinctus* (Nine-banded armadillo)

Order Lagomorpha (Rabbits, hares, and pikas)
        75400=Lagomorpha family indeterminate

    Family Ochotonidae
        75450=*Ochotona* sp.
        75500=*Ochotona princeps* (Pika)
        75550=*Ochotona collaris* (Collared pika)

    Family Leporidae
        75600=Leporidae genus indeterminate
        75650=*Lepus* sp.
        75700=*Lepus alleni* (Antelope jackrabbit)
        75750=*Lepus americanus* (Snowshoe hare)
        75800=*Lepus arcticus* (Arctic hare)
        75850=*Lepus californicus* (California or black-tailed
            jackrabbit)

75900=*Lepus callotis* (White-sided jackrabbit)
75950=*Lepus capensis* (Cape or European hare)
76000=*Lepus insularis* (Black Jack rabbit)
76050=*Lepus othus* (Tundra or Alaskan hare)
76100=*Lepus timidus* (Northern hare)
76150=*Lepus townsendii* (White-tailed jackrabbit)
76200=*Sylvilagus* sp.
76250=*Sylvilagus audubonii* (Audubon or desert cottontail rabbit)
76300=*Sylvilagus aquaticus* (Swamp rabbit)
76350=*Sylvilagus bachmani* (Brush rabbit)
76400=*Sylvilagus floridanus* (Eastern cottontail rabbit)
76450=*Sylvilagus idahoensis* (Pygmy rabbit)
76500=*Sylvilagus nuttallii* (Nuttall's cottontail)
76550=*Sylvilagus palustris* (Marsh rabbit)
76600=*Sylvilagus robustus* (Davis Mountains cottontail rabbit)
76650=*Sylvilagus transitionalis* (New England cottontail)
76700=*Oryctolagus cuniculus* (Domestic rabbit)

Order Rodentia (Rodents)
76750=Rodentia family indeterminate
76760=Small rodent (Mouse-sized)
76770=Medium rodent (Rat/gopher-sized)
76780=Large rodent (Muskrat/beaver-sized)

Family Aplodontidae
76800=*Aplodontia rufa* (Mountain beaver)

Family Sciuridae (Squirrels)
76850=Sciuridae genus indeterminate
76900=*Tamias striatus* (Eastern chipmunk)
76950=*Eutamias* sp.
77000=*Eutamias alpinus* (Alpine chipmunk)
77050=*Eutamias amoenus* (Yellow-pine chipmunk)
77100=*Eutamias canipes* (Gray-footed chipmunk)
77150=*Eutamias cinereicollis* (Gray-collared chipmunk)
77200=*Eutamias dorsalis* (Cliff chipmunk)
77250=*Eutamias merriami* (Merriam's chipmunk)
77300=*Eutamias minimus* (Least chipmunk)
77350=*Eutamias obscurus* (Baja California chipmunk)
77400=*Eutamias palmeri* (Charleston mountain chipmunk)
77450=*Eutamias panamintinus* (Panamint chipmunk)
77500=*Eutamias quadrimaculatus* (Long-eared chipmunk)
77550=*Eutamias quadrivittatus* (Colorado chipmunk)
77600=*Eutamias ruficaudus* (Red-tailed chipmunk)
77650=*Eutamias senex* (California chipmunk)
77700=*Eutamias siskiyou* (Siskiyou chipmunk)
77750=*Eutamias sonomae* (Sonoma chipmunk)
77800=*Eutamias speciosus* (Lodgepole chipmunk)
77850=*Eutamias townsendii* (Townsend's chipmunk)
77900=*Eutamias umbrinus* (Uinta chipmunk)
77950=*Marmota* sp.

78000=*Marmota caligata* (Hoary marmot)
78050=*Marmota flaviventris* (Yellow-bellied marmot)
78100=*Marmota monax* (Woodchuck or marmot)
78150=*Marmota olympus* (Olympic marmot)
78200=*Marmota vancouverensis* (Vancouver marmot)
78250=*Sciurus* sp.
78300=*Sciurus aberti* (Abert's squirrel)
78350=*Sciurus apache* (Apache fox squirrel)
78400=*Sciurus arizonensis* (Arizona gray squirrel)
78450=*Sciurus carolinensis* (Eastern gray squirrel)
78500=*Sciurus griseus* (Western gray squirrel)
78550=*Sciurus kaibabensis* (Kaibab squirrel)
78600=*Sciurus niger* (Fox squirrel)
78650=*Tamiasciurus* sp.
78700=*Tamiasciurus douglasii* (Douglas' squirrel)
78750=*Tamiasciurus hudsonicus* (Red squirrel)
78800=*Spermophilus* sp.
78850=*Spermophilus armatus* (Uinta ground squirrel)
78900=*Spermophilus atricapillus* (Baja California rock squirrel)
78950=*Spermophilus beecheyi* (California ground squirrel)
79000=*Spermophilus beldingi* (Belding's ground squirrel)
79050=*Spermophilus brunneus* (Idaho ground squirrel)
79100=*Spermophilus columbianus* (Columbian ground squirrel)
79150=*Spermophilus franklinii* (Franklin's ground squirrel)
79200=*Spermophilus lateralis* (Golden-mantled ground squirrel)
79250=*Spermophilus mexicanus* (Mexican ground squirrel)
79300=*Spermophilus mohavensis* (Mohave ground squirrel)
79350=*Spermophilus parryii* (Arctic ground squirrel)
79400=*Spermophilus richardsonii* (Richardson's ground squirrel)
79450=*Spermophilus saturatus* (Cascade Golden-mantled ground squirrel)
79500=*Spermophilus spilosoma* (Spotted ground squirrel)
79550=*Spermophilus tereticaudus* (Round-tailed ground squirrel)
79600=*Spermophilus townsendii* (Townsend's ground squirrel)
79650=*Spermophilus tridecemlineatus* (Thirteen-lined ground squirrel)
79700=*Spermophilus variegatus* (Rock squirrel)
79750=*Spermophilus washingtoni* (Washington ground squirrel)
79800=*Cynomys* sp.
79850=*Cynomys gunnisoni* (Gunnison's prairie dog)
79900=*Cynomys ludovicianus* (Black-tailed prairie dog)
79950=*Cynomys leucurus* (White-tailed prairie dog)
80000=*Ammospermophilus* sp.

80050=*Ammospermophilus harrisii* (Harris' antelope squirrel)
80100=*Ammospermophilus interpres* (Texas antelope ground squirrel)
80150=*Ammospermophilus insularis* (Espiritu Santo Island Antelope squirrel)
80200=*Ammospermophilus leucurus* (White-tailed antelope squirrel)
80250=*Ammospermophilus nelsoni* (San Joaquin antelope squirrel)
80300=*Glaucomys* sp.
80350=*Glaucomys sabrinus* (Northern flying squirrel)
80400=*Glaucomys volans* (Eastern flying squirrel)

Family Geomyidae (Pocket gophers)
80450=Geomyidae genus indeterminate
80500=*Geomys* sp. (*=no post-cranial differences)
80550=*Geomys arenarius* (Desert pocket gopher)
80600=*Geomys attwateri* (Attwater's pocket gopher*)
80650=*Geomys breviceps* (Louisiana pocket gopher*)
80700=*Geomys bursarius* (Plains pocket gopher*)
80750=*Geomys colonus* (Colonial pocket gopher)
80800=*Geomys cumberlandius* (Cumberland Island pocket gopher)
80850=*Geomys fontanelus* (Sherman's pocket gopher)
80900=*Geomys personatus* (South Texas pocket gopher)
80950=*Geomys pinetis* (Southeastern pocket gopher)
81000=*Pappogeomys castanops* (Yellow-faced pocket gopher)
81050=*Thomomys* sp.
81100=*Thomomys baileyi* (Bailey pocket gopher)
81150=*Thomomys bottae* (Botta pocket gopher)
81200=*Thomomys bulbivorus* (Camas pocket gopher)
81250=*Thomomys mazama* (Western pocket gopher)
81300=*Thomomys monticola* (Mountain pocket gopher)
81350=*Thomomys talpoides* (Northern pocket gopher)
81400=*Thomomys townsendii* (Townsend's pocket gopher)
81450=*Thomomys umbrinus* (Southern pocket gopher)

Family Heteromyidae (Pocket mice and kangaroo rats)
81500=Heteromyidae genus indeterminate
81550=*Liomys irroratus* (Mexican spiny pocket mouse)
81600=*Perognathus* sp.
81650=*Perognathus alticolus* (White-eared pocket mouse)
81700=*Perognathus amplus* (Arizona pocket mouse)
81750=*Perognathus apache* (Apache pocket mouse)
81800=*Perognathus arenarius* (Little desert pocket mouse)
81850=*Perognathus baileyi* (Bailey's pocket mouse)
81900=*Perognathus californicus* (California pocket mouse)
81950=*Perognathus dalquesti* (Dalquest's pocket mouse)
82000=*Perognathus fallax* (San Diego pocket mouse)
82050=*Perognathus fasciatus* (Olive-backed pocket mouse)

82100=*Perognathus flavescens* (Plains pocket mouse)
82150=*Perognathus flavus* (Baird or silky pocket mouse)
82200=*Perognathus formosus* (Long-tailed pocket mouse)
82250=*Perognathus hispidus* (Hispid pocket mouse)
82300=*Perognathus inornatus* (San Joaquin pocket mouse)
82350=*Perognathus intermedius* (Intermediate or rock pocket mouse)
82400=*Perognathus longimembris* (Little pocket mouse)
82450=*Perognathus merriami* (Merriam pocket mouse)
82500=*Perognathus nelsoni* (Nelson pocket mouse)
82550=*Perognathus parvus* (Great Basin pocket mouse)
82600=*Perognathus penicillatus* (Desert pocket mouse)
82650=*Perognathus spinatus* (Spiny pocket mouse)
82700=*Perognathus xanthonotus* (Walker pass pocket mouse)
82750=*Microdipodops* sp.
82800=*Microdipodops megacephalus* (Dark kangaroo mouse)
82850=*Microdipodops pallidus* (Pale kangaroo mouse)
82900=*Dipodomys* sp.
82950=*Dipodomys agilis* (Agile kangaroo rat)
83000=*Dipodomys compactus* (Padre Island kangaroo rat)
83050=*Dipodomys deserti* (Desert kangaroo rat)
83100=*Dipodomys elator* (Texas kangaroo rat)
83150=*Dipodomys elephantinus* (Big-eared kangaroo rat)
83200=*Dipodomys gravipes* (San Quintin kangaroo rat)
83250=*Dipodomys heermanni* (Heermann's kangaroo rat)
83300=*Dipodomys ingens* (Giant kangaroo rat)
83350=*Dipodomys insularis* (San Jose Island kangaroo rat)
83400=*Dipodomys margaritae* (Margarita Island kangaroo rat)
83450=*Dipodomys microps* (Chisel-toothed kangaroo rat)
83500=*Dipodomys merriami* (Merriam kangaroo rat)
83550=*Dipodomys nitratoides* (Fresno kangaroo rat)
83600=*Dipodomys ordii* (Ord kangaroo rat)
83650=*Dipodomys panamintinus* (Panamint kangaroo rat)
83700=*Dipodomys paralius* (Santa Catarina kangaroo rat)
83750=*Dipodomys peninsularis* (Baja California kangaroo rat)
83800=*Dipodomys spectabilis* (Banner-tailed kangaroo rat)
83850=*Dipodomys stephensi* (Stephen's kangaroo rat)
83900=*Dipodomys venustus* (Santa Cruz kangaroo rat)

Family Castoridae
83950=*Castor canadensis* (Beaver)

Family Cricetidae (New World rats and mice)
  84000=Cricetidae genus indeterminate
  84010=Small Cricetid rodent (Mouse-sized)
  84020=Medium Cricetid rodent (Rat-sized)
  84030=Large Cricetid Rodent (Muskrat-sized)
  84050=*Onychomys* sp.
  84100=*Onychomys leucogaster* (Short-tailed or northern grasshopper mouse)
  84150=*Onychomys torridus* (Long-tailed or southern grasshopper mouse)
  84200=*Reithrodontomys* sp.
  84250=*Reithrodontomys fulvescens* (Fulvous harvest mouse)
  84300=*Reithrodontomys humulis* (Dwarf harvest mouse)
  84350=*Reithrodontomys megalotis* (Western harvest mouse)
  84400=*Reithrodontomys montanus* (Plains harvest mouse)
  84450=*Reithrodontomys raviventris* (Salt-march harvest mouse)
  84500=*Baiomys taylori* (Northern pygmy mouse)
  84550=*Peromyscus* sp.
  84600=*Peromyscus attwateri* (Texas mouse)
  84650=*Peromyscus boylii* (Brush mouse)
  84700=*Peromyscus californicus* (California mouse)
  84750=*Peromyscus caniceps* (Burt's deer mouse)
  84800=*Peromyscus comanche* (Palo Duro mouse)
  84850=*Peromyscus crinitus* (Canyon Mouse)
  84900=*Peromyscus dickeyi* (Dickey's deer mouse)
  84950=*Peromyscus difficilis* (Rock mouse)
  85000=*Peromyscus eremicus* (Cactus mouse)
  85050=*Peromyscus eva* (Eva's desert mouse)
  85100=*Peromyscus floridanus* (Florida or gopher mouse)
  85150=*Peromyscus gossypinus* (Cotton mouse)
  85200=*Peromyscus guardia* (Angel Island mouse)
  85250=*Peromyscus interparietalis* (San Lorenzo deer mouse)
  85300=*Peromyscus leucopus* (White-footed mouse)
  85350=*Peromyscus maniculatus* (Deer mouse)
  85400=*Peromyscus merriami* (Merriam's mouse)
  85450=*Peromyscus nuttalli* (Golden mouse)
  85500=*Peromyscus pectoralis* (Encinal or white-ankled mouse)
  85550=*Peromyscus polionotus* (Oldfield mouse)
  85600=*Peromyscus pseudocrinitis* (False canyon mouse)
  85650=*Peromyscus sejugis* (Santa Cruz Island mouse)
  85700=*Peromyscus sitkensis* (Sitka mouse)
  85750=*Peromyscus slevini* (Slevin's mouse)
  85800=*Peromyscus stephani* (San Esteban Island mouse)
  85850=*Peromyscus truei* (Pinon mouse)
  85900=*Oryzomys* sp.
  85950=*Oryzomys argentatus* (Key rice rat)
  86000=*Oryzomys couesi* (Coues rice rat)
  86050=*Oryzomys palustris* (Marsh rice rat)
  86100=*Ochrotomys nuttalli* (Golden mouse)
  86150=*Sigmodon* sp.

86200=*Sigmodon fulviventer* (Tawny-bellied cotton rat)
86250=*Sigmodon hispidus* (Hispid cotton rat)
86300=*Sigmodon minimus* (Least cotton rat)
86350=*Sigmodon ochrognathus* (Yellow-nosed cotton rat)
86400=*Neotoma* sp.
86450=*Neotoma albigula* (White-throated wood rat)
86500=*Neotoma anthonyi* (Anthony's wood rat)
86550=*Neotoma bunkeri* (Bunker's wood rat)
86600=*Neotoma cinerea* (Bushy-tailed wood rat)
86650=*Neotoma floridana* (Florida wood rat)
86700=*Neotoma fuscipes* (Dusky-footed wood rat)
86750=*Neotoma lepida* (Desert wood rat)
86800=*Neotoma martinensis* (San Martin Is. wood rat)
86850=*Neotoma mexicana* (Mexican wood rat)
86900=*Neotoma micropus* (Gray or S. plains wood rat)
86950=*Neotoma stephensi* (Stephen's wood rat)
87000=*Clethrionomys* sp.
87050=*Clethrionomys californicus* (California red-backed vole)
87100=*Clethrionomys gapperi* (Gapper's red-backed mouse)
87150=*Clethrionomys rutilus* (Tundra red-backed vole)
87200=*Phenacomys* sp.
87250=*Phenacomys albipes* (White-footed vole)
87300=*Phenacomys intermedius* (Heather vole)
87350=*Phenacomys longicaudus* (Red tree vole)
87400=*Microtus* sp.
87450=*Microtus abbreviatus* (Insular vole)
87500=*Microtus breweri* (Beach vole)
87550=*Microtus californicus* (California vole)
87600=*Microtus chrotorrhinus* (Rock vole)
87650=*Microtus coronarius* (Coronation Island vole)
87700=*Microtus longicaudus* (Long-tailed vole)
87750=*Microtus mexicanus* (Mexican vole)
87800=*Microtus miurus* (Alaska or singing vole)
87850=*Microtus montanus* (Montane vole)
87900=*Microtus nesophilus* (Gull Island vole)
87950=*Microtus ochrogaster* (Prairie vole)
88000=*Microtus oeconomus* (Tundra vole)
88050=*Microtus oregoni* (Creeping vole)
88100=*Microtus pennsylvanicus* (Meadow vole)
88150=*Microtus pinetorum* (Pine vole)
88200=*Microtus richardsoni* (Richardson's vole)
88250=*Microtus townsendii* (Townsend's vole)
88300=*Microtus xanthognathus* (Yellow-cheeked vole)
88350=*Arvicola richardsoni* (Water vole)
88400=*Lagurus curtatus* (Sagebrush vole)
88450=*Neofiber alleni* (Round-tailed muskrat)
88500=*Ondatra zibethicus* (Muskrat)
88550=*Synaptomys* sp.
88600=*Synaptomys cooperi* (Southern bog lemming)
88650=*Synaptomys borealis* (Northern bog lemming)
88700=*Dicrostonyx* sp.
88750=*Dicrostonyx exsul* (St. Lawrence Island collared lemming)

88800=*Dicrostonyx groenlandicus* (Greenland collared lemming)
88850=*Dicrostonyx hudsonius* (Hudson Bay collared lemming)
88900=*Dicrostonyx torquatus* (Collared lemming)
88950=*Lemmus sibiricus* (Brown lemming)

Family Muridae
89000=Muridae genus indeterminate
89050=*Mus musculus* (House mouse)
89100=*Rattus* sp.
89150=*Rattus norvegicus* (Norway rat)
89200=*Rattus rattus* (Roof or black rat)

Family Zapodidae
89250=Zapodidae genus indeterminate
89300=*Zapus* sp.
89350=*Zapus hudsonius* (Meadow jumping mouse)
89400=*Zapus princeps* (Western jumping mouse)
89450=*Zapus trinotatus* (Pacific jumping mouse)
89500=*Napaeozapus insignis* (Woodland jumping mouse)

Family Erethizontidae
89550=*Erethizon dorsatum* (Porcupine)

Family Myocastoridae
89600=*Myocastor coypus* (Nutria)

Order Carnivora (Carnivores)
89650=Carnivora family indeterminate

Family Ursidae (Bears)
89700=*Ursus* sp.
89750=*Ursus americanus* (Black bear)
89800=*Ursus arctos* (Grizzly bear)
89850=*Ursus maritimus* (Polar bear)
89900=*Ursus middendorffi* (Alaskan brown bear)

Family Procyonidae (Raccoons and relatives)
89950=Procyonidae genus indeterminate
90000=*Procyon lotor* (Raccoon)
90050=*Bassariscus astutus* (Ringtail)
90100=*Nasua nasua* (Coati)

Family Mustelidae (Weasels and relatives)
90150=Mustelidae genus indeterminate
90200=*Mustela* sp.
90250=*Mustela erminea* (Ermine)
90300=*Mustela frenata* (Long-tailed weasel)
90350=*Mustela macrodon* (Sea mink)
90400=*Mustela nigripes* (Black-footed ferret)
90450=*Mustela nivalis* (Least weasel)
90500=*Mustela vison* (Mink)
90550=*Martes* sp.
90600=*Martes americana* (Marten)
90650=*Martes pennanti* (Fisher)
90700=*Spilogale* sp.

90750=*Spilogale gracilis* (Western spotted skunk)
90800=*Spilogale putorius* (Eastern spotted skunk)
90850=*Mephitis* sp.
90900=*Mephitis macroura* (Hooded skunk)
90950=*Mephitis mephitis* (Striped skunk)
91000=*Conepatus* sp.
91050=*Conepatus mesoleucus* (Hog-nosed skunk)
91100=*Conepatus leuconotus* (Gulf coast hog-nosed skunk)
91150=*Taxidea taxus* (Badger)
91200=*Lutra canadensis* (River otter)
91250=*Enhydra lutris* (Sea otter)
91300=*Gulo luscus* (Wolverine)

Family Canidae (Dogs and relatives)
91350=Canidae genus indeterminate
91400=*Vulpes* sp.
91450=*Vulpes macrotis* (Desert fox)
91500=*Vulpes velox* (Swift kit fox)
91550=*Vulpes vulpes* (Red fox)
91600=*Alopex lagopus* (Arctic fox)
91650=*Urocyon* sp.
91700=*Urocyon cinereoargenteus* (Gray fox)
91750=*Urocyon littoralis* (Island gray fox)
91800=*Canis* sp.
91850=*Canis familiaris* (Domestic dog)
91900=*Canis latrans* (Coyote)
91950=*Canis lupus* (Gray wolf)
92000=*Canis rufus* (Red wolf)

Family Felidae (Cats)
92050=*Felis* sp.
92100=*Felis concolor* (Cougar)
92150=*Felis domesticus* (Domestic cat)
92200=*Felis lynx* (Lynx)
92250=*Felis onca* (Jaguar)
92300=*Felis pardalis* (Ocelot)
92350=*Felis rufus* (Bobcat)
92400=*Felis weidii* (Margay)
92450=*Felis yagouaroundi* (Jaguarundi)

Order Pinnipedia (Seals and allies)
92500=Pinnipedia family indeterminate

Family Otariidae
92550=Otariidae genus indeterminate
92600=*Callorhinus ursinus* (Northern fur seal)
92650=*Arctocephalus philippii* (Guadalupe fur seal)
92700=*Eumetopias jubatus* (Northern sea lion)
92750=*Zalophus californianus* (California sea lion)

Family Odobenidae (Walrus)
92800=*Odobenus rosmarus* (Walrus)

Family Phocidae (Hair seals)
92850=Phocidae genus indeterminate
92900=*Phoca* sp.

92950=*Phoca fasciata* (Ribbon seal)
93000=*Phoca groenlandicus* (Harp seal)
93050=*Phoca hispida* (Ringed seal)
93100=*Phoca vitulina* (Harbor seal)
93150=*Erignathus barbatus* (Bearded seal)
93200=*Halichoerus grypus* (Gray seal)
93250=*Monachus tropicalis* (West Indian monk seal)
93300=*Cystophora cristata* (Hooded seal)
93350=*Mirounga angustirostris* (Northern elephant seal)

Order Sirenia
93400=*Trichechus manatus* (Manatee or sea cow)

Order Artiodactyla (Even-toed ungulates)
93450=Artiodactyla family indeterminate
93460=Small Artiodactyl (Peccary, domestic goat-sized)
93470=Medium Artiodactyl (Deer, pronghorn-sized)
93480=Large Artiodactyl (Elk, caribou, bison-sized)

Family Tayassuidae
93500=*Dicotyles tajacu* (Collared peccary: javelina)

Family Suidae
93550=*Sus scrofa* (Pig)

93600=*Dicotyles/Sus* genus indeterminate

Family Cervidae (Deer and relatives)
93650=Cervidae genus indeterminate
93700=*Cervus* sp.
93750=*Cervus elephus* (Elk or Wapiti)
93800=*Cervus dama* (Fallow deer)
93850=*Cervus merriami* (Merriam's elk)
93900=*Alces alces* (Moose)
93950=*Rangifer* sp.
94000=*Rangifer arcticus* (Barren ground caribou)
94050=*Rangifer caribou* (Woodland caribou)
94100=*Rangifer tarandus* (Greenland caribou)
94150=*Axis axis* (Axis deer)
94200=*Odocoileus* sp.
94250=*Odocoileus hemionus* (Mule deer)
94300=*Odocoileus virginianus* (White-tailed deer)

Family Antilocapridae
94350=*Antilocapra americana* (Pronghorn antelope)

94400=*Antilocapra/Odocoileus* genus indeterminate

Family Bovidae (Cattle and relatives)
94450=Bovidae genus indeterminate
94500=*Bison* sp. indeterminate
94550=*Bison bison* (Plains bison)
94570=*Bos* sp.
94580=*Bos indicus* (Indian cow)
94600=*Bos taurus* (Cow)
94650=*Bos/Bison* genus indeterminate

      94700=*Oreamnos americanus* (Mountain goat)
      94750=*Ovibos moschatus* (Muskox)
      94800=*Capra hircus* (Goat)
      94850=*Ovis* sp.
      94900=*Ovis aries* (Sheep)
      94950=*Ovis dalli* (Dall sheep)
      95000=*Ovis canadensis* (Bighorn or mountain sheep)
      95050=*Ovis/Capra* genus indeterminate
      95100=*Ammotragus lervia* (Barbary sheep: aoudad)

    Family Camelidae (Camels)
      95150=*Camelus* sp.

Order Perissodactyla (Odd-toed ungulates)
    Family Equidae
      95200=*Equus* sp.
      95250=*Equus caballas* (Modern horse)
      95300=*Equus asinus* (Donkey)

Order Mysticeti (Baleen whales)
      95350=Mysticeti family indeterminate

    Family Eschrichtidae
      95400=*Eschrichtius gibbosus* (Gray whale)

    Family Baleanopteridae (Finback whales)
      95450=Baleanopteridae genus indeterminate
      95500=*Balaenoptera* sp.
      95550=*Balaenoptera acutorostrata* (Piked whale)
      95600=*Balaenoptera borealis* (Rorqual or Sei whale)
      95650=*Balaenoptera edeni* (Bryle's whale)
      95700=*Balaenoptera musculus* (Blue whale)
      95750=*Balaenoptera physalus* (Common finback whale)
      95800=*Megaptera novaeangliae* (Humpback whale)

    Family Balaenidae (Right and bowhead whales)
      95850=*Balaena* sp.
      95900=*Balaena glacialis* (Right whale)
      95950=*Balaena mysticetus* (Bowhead whale)

Order Odontoceti (Toothed whales)
      96000=Odontoceti family indeterminate

    Family Physeteridae (Sperm whales)
      96050=*Physeter catodon* (Sperm whale)

    Family Kogiidae (Pygmy sperm whales)
      96100=*Kogia* sp.
      96150=*Kogia breviceps* (Pygmy sperm whale)
      96200=*Kogia simus* (Dwarf sperm whale)

    Family Ziphiidae (Beaked whales)
      96250=Ziphiidae genus indeterminate
      96300=*Ziphius cavirostris* (Goose-beaked whale)
      96350=*Berardius bairdii* (Baird beaked whale)
      96400=*Mesoplodon* sp.
      96450=*Mesoplodon bidens* (Sowerby beaked whale)

96500=*Mesoplodon carlhubbsi* (Archbeak whale)
96550=*Mesoplodon densirostris* (Atlantic beaked whale)
96600=*Mesoplodon europaeus* (Gulf-stream beaked whale)
96650=*Mesoplodon ginkgodens* (Japanese beaked whale)
96700=*Mesoplodon mirus* (True's beaked whale)
96750=*Mesoplodon stejnegeri* (Pacific beaked whale)
96800=*Hyperoodon ampullatus* (Bottlenose whale)

Family Monodontidae
96850=Monodontidae genus indeterminate
96900=*Delphinapterus leucas* (White whale)
96950=*Monodon monoceros* (Narwhal)

Family Delphinidae (Dolphins)
97000=Delphinidae genus indeterminate
97050=*Delphinus delphis* (Common dolphin)
97100=*Steno bredanensis* (Rough-toothed whale or dolphin)
97150=*Stenella* sp.
97200=*Stenella coeruleoalba* (Striped dolphin)
97250=*Stenella frontalis* (Bridled dolphin)
97300=*Stenella longirostris* (Longbeak dolphin)
97350=*Stenella plagiodon* (Spotted dolphin)
97400=*Tursiops* sp.
97450=*Tursiops gilli* (Pacific bottle-nosed dolphin)
97500=*Tursiops truncatus* (Bottle-nosed dolphin)
97550=*Lissodelphis borealis* (Right whale dolphin)
97600=*Lagenorhynchus* sp.
97650=*Lagenorhynchus acutus* (Atlantic white-sided dolphin)
97700=*Lagenorhynchus albirostris* (Whitebeak dolphin)
97750=*Lagenorhynchus obliquidens* (Pacific white-sided dolphin)
97800=*Globicephala* sp.
97850=*Globicephala macrorhyncha* (Short-finned blackfish)
97900=*Globicephala melaena* (Blackfish or pilot whale)
97950=*Orcinus orca* (Killer whale)
98000=*Pseudorca crassidens* (False killer whale)
98050=*Feresa attenuata* (Pygmy killer whale)
98100=*Grampus griseus* (Grampus or Risso dolphin)
98150=*Phocoena phocoena* (Harbor porpoise)
98200=*Phocoenoides dalli* (Dall porpoise)

Order Primates
99900=*Homo sapiens sapiens* (Modern human)

APPENDIX II

dBASE FACS SUPPORT PROGRAMS AND PROCEDURES

by David L. Carlson and Brian S. Shaffer

INTRODUCTION

Three programs have been developed to aid in the analysis and reporting of faunal data associated with this coding system. Each of these programs, the CHECK program, the XTOT program, and the LINKFILE programs are listed here along with instructions for use. Also included is a set of procedures for condensing a dBase file down to the fewest number of unique lines. Another set of procedures has been provided to link associated databases, such as a provenience database and faunal database, based on a common field(s) in the two databases. In this way, additional provenience information can be added quickly without having to add it to each line of data individually. The use of these programs and procedures requires a fundamental knowledge of dBase for creating and executing programs and dBase files. DISCLAIMER: Every data set manipulated electronically is subject to the hazards associated with computer use. DO NOT ATTEMPT TO USE ANY OF THESE PROGRAMS OR PROCEDURES WITHOUT FIRST MAKING AT LEAST ONE BACKUP OF YOUR DATA. Although these programs have been tested extensively and we currently use them in our analyses, we assume no risk for their use by other researchers.

CHECK PROGRAM

The CHECK program (Table 2) is designed to locate many of the illogical errors that may result due to miscoding or erroneous data entry. This program will NOT replace the process of verifying the data that has been entered. We suggest that researchers visually inspect their data to ensure that what was recorded on the hard copy pages is what was encoded into the computer. The CHECK program, however, is designed to locate illogical errors that may have been missed in the verification process, as well as additional errors that result from miscoding. Table 3 provides a listing of abbreviations for each field name that must be used for the CHECK program. Table 4 lists the 19 individual error categories, representing 120 individual checks per record (line of data), that are currently included in the program.

The CHECK program works by locating relational errors between fields or a number range within a field. As an example, for a taxon field coded as mammal, the program checks to ensure that the element field has not been coded using numbers associated with non-mammalian elements. Another example would be between the element and portion of element category. Once again, the program checks to see that the two are congruent with each other. If not, the line is flagged with an error code. Note: for a database to be checked for errors by the CHECK program, the field names in the database file must contain exactly the names listed in Table 3.

Table 2. CHECK Program.

```
*CHECK PROGRAM

CLEAR
USE
fname=SPACE(25)

? "FACS CHECK PROGRAM"

@10,10 SAY "What is the name of the Database you want to check?"
@11,25 GET fname
READ
fname=Trim(ltrim(UPPER(fname)))

USE &fname

IF TYPE("ERROR") = "U"
    ?
    ? "  THE ERROR FIELD HAS NOT BEEN INCLUDED IN THIS FILE."
    ? "    PLEASE WAIT WHILE THE DATABASE IS MODIFIED."
    COPY TO TEMPFLE STRUCTURE EXTENDED
    USE TEMPFLE
    APPEND BLANK
    REPLACE FIELD_NAME WITH "ERROR", FIELD_TYPE WITH "N", FIELD_LEN WITH 2, FIELD_DEC WITH 0
    USE
    CREATE TEMPXXX FROM TEMPFLE
    ERASE TEMPFLE.DBF
    ?
    ? "DATABASE MODIFIED.  NOW APPENDING RECORDS WITH ERROR FIELD."
    SET TALK ON
    APPEND FROM &fname
    *SET TALK OFF
    USE
    dbname = fname+".DBF"
    bkname = fname+".BAK"
    RENAME &dbname TO &bkname
    RENAME TEMPXXX.DBF TO &dbname
    USE &fname
ENDIF

CLEAR
```

Table 2. Continued.

*CHECKING FOR RECORDS WITH 0 QUANTITY
REPL ALL ERROR WITH 1 FOR QTY=0

*FISH
REPL ALL ERROR WITH 5 FOR TXN>99 .AND. TXN<30000 .AND. (EL>10 .AND. EL<501 .OR. EL>650)

*AMPHIBIANS
REPL ALL ERROR WITH 5 FOR TXN>30000 .AND. TXN<40000 .AND. (EL>500 .AND. EL<650 .OR. EL>700)

*REPTILES
REPL ALL ERROR WITH 5 FOR TXN>40000 .AND. TXN<50000 .AND. (EL>500 .AND. EL<700 .OR. EL>800)

*BIRDS
REPL ALL ERROR WITH 5 FOR TXN>50000 .AND. TXN<70000 .AND. (EL>900 .OR. EL>500 .AND. EL<801 .OR. EL>401 .AND. EL<424)
REPL ALL ERROR WITH 5 FOR TXN>50000 .AND. TXN<70000 .AND. EL>426 .AND. EL<447

*MAMMALS
REPL ALL ERROR WITH 5 FOR TXN>70000 .AND. EL>500 .AND. EL<900

*INDETERMINATE
REPL ALL ERROR WITH 10 FOR EL=0 .AND. PE>0 .AND. PE<901

*SKELETON
REPL ALL ERROR WITH 10 FOR EL=1 .AND. PE#901

*SKULL
REPL ALL ERROR WITH 10 FOR EL=10 .AND. (PE=0 .OR. PE>99 .AND. PE<999)

*MANDIBLE
REPL ALL ERROR WITH 10 FOR EL=20 .AND. (PE>1 .AND. PE<101 .OR. PE>200 .AND. PE<999)

*TOOTH
REPL ALL ERROR WITH 10 FOR EL>29 .AND. EL<33 .AND. (PE>1 .AND. PE<201 .OR. PE>300)

*VERTEBRA
REPL ALL ERROR WITH 10 FOR EL>39 .AND. EL<111 .AND. (PE>1 .AND. PE>301 .OR. PE>399 .AND. PE<999)

*STERNUM
REPL ALL ERROR WITH 10 FOR EL=130 .AND. (PE>1 .AND. PE<400 .OR. PE>450 .AND. PE<999)

Table 2. Continued.

*RIB
REPL ALL ERROR WITH 10 FOR EL>139 .AND. EL<149 .AND. (PE>1 .AND. PE<450 .OR. PE>499 .AND. PE<999)

*CLAVICLE AND CORACOID
REPL ALL ERROR WITH 10 FOR EL>149 .AND. EL<161 .AND. (PE>1 .AND. PE<601 .OR. PE>699 .AND. PE<999)

*SCAPULA
REPL ALL ERROR WITH 10 FOR EL=170 .AND. (PE>1 .AND. PE<501 .OR. PE>599 .AND. PE<999)

*UPPER LONG BONES
REPL ALL ERROR WITH 10 FOR EL>179 .AND. EL<201 .AND. (PE>1 .AND. PE<601 .OR. PE>699 .AND. PE<999)

*FEMUR
REPL ALL ERROR WITH 10 FOR EL=220 .AND. (PE>1 .AND. PE<601 .OR. PE>699 .AND. PE<999)

*TIBIA
REPL ALL ERROR WITH 10 FOR EL=240 .AND. (PE>1 .AND. PE<601 .OR. PE>699 .AND. PE<999)

*FIBULA
REPL ALL ERROR WITH 10 FOR EL=250 .AND. (PE>1 .AND. PE<601 .OR. PE>699 .AND. PE<999)

*METAPODIALS AND PHALANGES--ALL
REPL ALL ERROR WITH 10 FOR EL>259 .AND. EL<390 .AND. (PE>1 .AND. PE<601 .OR. PE>699 .AND. PE<999)

*PELVIS
REPL ALL ERROR WITH 10 FOR EL=210 .AND. (PE>1 .AND. PE<701 .OR. PE>799 .AND. PE<999)

*PATELLA
REPL ALL ERROR WITH 10 FOR EL=230 .AND. (PE>1 .AND. PE<901)

*DISTAL FIBULA
REPL ALL ERROR WITH 10 FOR EL=251 .AND. (PE>1 .AND. PE<901)

*CARPALS, TARSALS, AND SESAMOIDS
REPL ALL ERROR WITH 10 FOR EL>399 .AND. EL<490 .AND. (PE>1 .AND. PE<901)

*LONG BONE INDETERMINATE
REPL ALL ERROR WITH 10 FOR EL=490 .AND. (PE>1 .AND. PE<601 .OR. PE>699 .AND. PE<999)

*COMPACT BONE INDETERMINATE, EPIPHYSES, AND FLAT BONE INDETERMINATE
REPL ALL ERROR WITH 10 FOR EL>491 .AND. EL<497 .AND. (PE>1 .AND. PE<901)

Table 2. Continued.

*ANTLER
REPL ALL ERROR WITH 10 FOR EL=901 .AND. (PE>1 .AND. PE<801 .OR. PE>830 .AND. PE<999)

*HORN
REPL ALL ERROR WITH 10 FOR EL=910 .AND. (PE>1 .AND. PE<841 .OR. PE>860 .AND. PE<999)

*FISH VERTEBRA
REPL ALL ERROR WITH 10 FOR EL>513 .AND. EL<527 .AND. (PE>1 .AND. PE<301 .OR. PE>399 .AND. PE<999)

*BIRD VERTEBRA
REPL ALL ERROR WITH 10 FOR EL=830 .AND. (PE>1 .AND. PE<301 .OR. PE>399) .OR. EL=840 .AND. (PE>1 .AND. PE<301 .OR. PE>399)

*PORTION CODED AS MOLAR AND ELEMENT NOT CODED AS PERMANENT TOOTH
REPL ALL ERROR WITH 15 FOR (PE>239 .AND. PE<248 .OR. PE>269 .AND. PE<278) .AND. (EL=31 .OR. EL=32)

*SIDE
REPL ALL ERROR WITH 20 FOR SD>3

*SIDE AND NON-AXIAL ELEMENTS
REPL ALL ERROR WITH 25 FOR (EL>10 .AND. EL<40 .OR. EL>131 .OR. EL<139 .AND. EL>210 .OR. EL>210 .AND. EL<505) .AND. SD>2
REPL ALL ERROR WITH 25 FOR (EL>650 .AND. EL<686 .OR. EL>706 .OR. EL>807 .AND. EL<826 .OR. EL>900 .AND. EL<920) .AND. SD>2
REPL ALL ERROR WITH 25 FOR (EL=705 .OR. EL=706 .AND. SD>2 .OR. EL=721 .OR. EL=936) .AND. SD>2
REPL ALL ERROR WITH 25 FOR (EL=940 .OR. EL>722 .AND. EL<727 .AND. EL=210 .AND. PE>701 .AND. PE<799 .AND. PE#790) .AND. SD>2

*SIDE AND AXIAL ELEMENTS
REPL ALL ERROR WITH 30 FOR (EL>39 .AND. EL<131 .OR. EL>513 .AND. EL<527 .AND. SD#3 .OR. EL=690 .OR. EL>702 .AND. EL<705) .AND. SD#3
REPL ALL ERROR WITH 30 FOR (EL>706 .AND. EL<709 .OR. EL=722 .OR. EL=740 .OR. EL=920 .OR. EL=930 .OR. EL=932) .AND. SD#3
REPL ALL ERROR WITH 30 FOR (EL=210 .AND. PE=790 .OR. EL=934 .OR. EL>800 .AND. EL<807 .OR. EL=210 .AND. PE=701) .AND. SD#3

*AGE
REPL ALL ERROR WITH 35 FOR AG>8

*AGE CRITERIA
REPL ALL ERROR WITH 40 FOR AC>65

*AGE & AGE CRITERIA
REPL ALL ERROR WITH 45 FOR AG>0 .AND. AC=0

*TOOTH TYPE AND AGE
REPL ALL ERROR WITH 50 FOR EL=31 .AND. AG>6

Table 2. Continued.

```
*TOOTH TYPE AND AGE CRITERIA
REPL ALL ERROR WITH 55 FOR EL=31 .AND. AC=37 .OR. EL=30 .AND. AC=36

*SEX
REPL ALL ERROR WITH 60 FOR SX>2

*SEX CRITERIA
REPL ALL ERROR WITH 65 FOR SC>7

*CHECKING ELEMENT AND SEX
REPL ALL ERROR WITH 70 FOR EL=930 .AND. SX#2 .OR. EL=920 .AND. SX#1

*CHECKING PORTION AND BREAKAGE
REPL ALL ERROR WITH 75 FOR PE>101 .AND. PE<200 .AND. PE#111 .AND. BK=0
REPL ALL ERROR WITH 75 FOR PE>301 .AND. PE<400 .AND. PE#315 .AND. PE#318 .AND. PE#321 .AND. BK=0
REPL ALL ERROR WITH 75 FOR (PE>451 .AND. PE<500 .AND. PE>501 .AND. PE<600 .OR. PE>603 .AND. PE<620 .OR. PE>629 .AND. PE<640) .AND. BK=0
REPL ALL ERROR WITH 75 FOR (PE>640 .AND. PE<670 .OR. PE>671 .AND. PE<701 .OR. PE>903 .OR. PE>801 .AND. PE<830) .AND. BK=0
REPL ALL ERROR WITH 75 FOR (PE>841 .AND. PE<860 .OR. PE>419 .AND. PE<450 .OR. PE=715 .OR. PE=720 .OR. PE=725) .AND. BK=0
REPL ALL ERROR WITH 75 FOR (PE=730 .OR. PE=745 .AND. BK=0 .OR. PE=749 .OR. PE=755) .AND. BK=0
REPL ALL ERROR WITH 75 FOR (PE=757 .OR. PE=759 .AND. BK=0 .OR. PE=765 .OR. PE=767 .OR. PE=769 .OR. PE=799) .AND. BK=0

*WEATHERING
REPL ALL ERROR WITH 80 FOR W>2

*BREAKAGE
REPL ALL ERROR WITH 85 FOR BK>8

*CUT
REPL ALL ERROR WITH 90 FOR CT>2

? "NUMBER OR RECORDS WITH ERRORS"
COUNT FOR ERROR>0
? ""

*ASK TO LIST TO THE SCREEN
ACCEPT "LIST PROBLEM RECORDS ON SCREEN? (Y/N)" TO YN
IF UPPER (YN)="Y"
   LIST ALL FOR ERROR>0
   WAIT
ENDIF
? ""
```

Table 2. Continued.

```
*ASK TO PRINT PROBLEMS
ACCEPT "PRINT PROBLEM RECORDS ON THE PRINTER? (Y/N)" TO YN

IF UPPER (YN)="Y"
    WAIT
      SET PRINT ON
        LIST ALL FOR ERROR>0
      SET PRINT OFF
ENDIF

? ""
*ASK TO ERASE ERROR CODES FROM DATABASE
ACCEPT "ERASE ERROR CODES FROM THE DATABASE? (Y/N)" TO YN
IF UPPER (YN)="Y"
    REPL ALL ERROR WITH 0 FOR ERROR>0
ENDIF
CLEAR ALL
CLEAR
```

Table 3. Field Name Abbreviations for the CHECK Program.

Abbreviation	Field Name
QTY	Quantity
TXN	Taxon
EL	Element
PE	Portion of Element
SD	Side
AC	Age Criteria
AG	Age
SX	Sex
SC	Sex Criteria
W	Weathering
BK	Breakage
G	Gnawing
CT	Cut

When the CHECK program is implemented (DO CHECK), the user is prompted for the file to be checked. The program next looks at your database file to see if the error field ("ERROR") has been included in the database. If the ERROR field is not present, the program can add the field to the end of the database if the database and CHECK program are in the same disk drive and directory. Do not use the CHECK program if the ERROR field is not in the data file and the CHECK program and data file are not both in the same disk drive and directory. This is because dBase will be unsuccessful in adding the ERROR field to the file and will have renamed the data file with a "BAK" extension. If the ERROR field is already present, it is not necessary for the CHECK program and database to be in the same drive or directory. Once the field has been added, the search for errors begins. Any field found with an error will have a number assigned to the "ERROR" column. The number assigned indicates the type of error located (Table 4). The error number must simply be compared to the error key (Table 4) to determine what type of error has occurred. However, since only one error can be recorded in the error field for each line of data, the last error encountered is the error that is recorded. Thus, we suggest examining the entire line of data that has been flagged, or running the CHECK program again after corrections have been made.

Once the data has been checked, the program will ask if the user would like to list just the lines with errors on the screen. The user also has the option of listing the errors on the printer. If the user chooses to list the lines with errors to the printer or screen, the lines will be listed in order of data entry, that is, by record number. The CHECK program lists or prints those fields that have an ERROR code greater than zero. Once each of these options has been completed, the CHECK program asks if the user would like errors in the error field to be reset to zero. If corrections were made and the CHECK program run again, without resetting the error field to zero, the same lines would appear as having errors because the ERROR field had not been reset. The ERROR field can be reset at any time by typing REPLACE ALL ERROR WITH 0 FOR ERROR>0.

Although North American taxonomy is listed with this system, the CHECK program will not be negated if taxonomies from other parts of the world are used, so long as the number ranges for each taxonomic class do not differ from those in this system.

Table 4. Error Key for CHECK Program.

Error # Error Description

Error #	Error Description
1	QUANTITY = 0
5	Discrepancy between TAXON & ELEMENT
10	Discrepancy between ELEMENT & PORTION OF ELEMENT
15	ELEMENT coded as deciduous tooth, PORTION OF ELEMENT coded as a molar
20	SIDE > 3 (Out of range for this category)
25	Discrepancy between SIDE & non-axial ELEMENTS should be denoted with a 2, 1, or 0.
30	Discrepancy between SIDE & axial ELEMENTS should be denoted with a 3.
35	AGE > 8 (Out of range for this category)
40	AGE CRITERIA > 65 (Out of range for this category)
45	Discrepancy between AGE & AGE CRITERIA
50	Deciduous tooth with improper AGE assignment
55	Discrepancy between tooth type and AGE CRITERIA
60	SEX > 3 (Out of range for this category)
65	SEX CRITERIA > 6 (Out of range for this category)
70	Discrepancy between ELEMENT & SEX
75	PORTION OF ELEMENT coded as a fragment, but BREAKAGE is coded as complete (0). The zero should be changed to a 1 or 2.
80	WEATHERING > 2 (Out of range for this category)
85	BREAKAGE > 8 (Out of range for this category)
90	CUT > 2 (Out of range for this category)

To reset the "ERROR" column, at the dot prompt type:
"REPLACE ALL ERROR WITH 0 FOR ERROR>0"

XTOT PROGRAM

The data sort program, XTOT (Table 5), is designed to condense a database file down to the fewest unique number of lines, based upon user-selected fields. This program is well adapted for generating taxon and element lists for estimating MNI or determining body portion representation. When XTOT is implemented (DO XTOT), it will ask for the file that the user would like to convert. The name of the database file, without the extension, should be entered into the box provided. Next, the program will ask for the number of variables that the user would like to select. These variables are the attribute or field names used in the file. Typically, the categories to be used will include taxon (TXN), element (EL), portion of element (PE), side (SD), and age (AG) for estimating MNI or body portion representations. Provenience may also be used. However, no more than eight variables may be used at one time due to the constraints of dBase. The program will then ask which field should be totaled. This should be the quantity (QTY) field. The program will then make a new database with just those fields specified by the user. The data will be sorted in ascending order by field. Each line is unique from the rest and is totaled.

The new database that has been created by the XTOT program has the same name as the original database, but with an "X" added to it. If the name of the database is shorter than eight characters, an "X" has been added to the end of the name. If the database file name is eight characters long, then an "X" will have replaced the eighth character. To avoid confusion, do not end any of the original data files with an "X." WARNING: The program does not first look to see if there is another file by the same name ending in "X." If there is, the old file will be replaced by the new file. Also, the names "TEMPSUM.DBF" OR "ORIGFILE.DBF" cannot be used to name database files. These file names are used, and later deleted by the XTOT program.

For XTOT to execute properly, the buffers and files setting within the CONFIG.SYS file of your computer must equal at least 10 each. Refer to your DOS manual for discussions of buffers and files within CONFIG.SYS. These settings are also required for the LINKFILE program.

Table 5. XTOT Program.

```
* XTOT.PRG

SET TALK OFF
CLEAR
nvar=0
origfile=SPAC(40)
@ 3,0 SAY "FACS CROSS TOTALING PROGRAM"
@ 5,0 SAY "Which file? " GET origfile
READ
origfile=UPPER(TRIM(origfile))
@ 7,0 SAY "How many variables? " GET nvar
READ
SELE 1
USE &origfile ALIA origfile
COPY STRU EXTENDED TO tempsum
SELE 2
USE tempsum
DELETE ALL
SELE 1
keyexp=""
* keyexp = list of vars delimited with plus signs

i=1
DO WHILE i<=nvar .AND. LEN(keyexp)<250
  nomvar=SPAC(10)
  @ 8+i, 0 SAY "Variable "+STR(i,2)+" field? " GET nomvar
  READ
  IF len(trim(nomvar))=0
    RETURN
  ENDIF
  SELE 2
  LOCATE FOR field_name=upper(nomvar)
  IF FOUND()
    RECALL
    IF field_type='N'

keyexp=keyexp+'STR('+nomvar+','+STR(field_len,3)+','+STR(field_dec,2)+')'
    ELSE
       keyexp=keyexp+nomvar
    ENDIF

    IF i<nvar
      keyexp=keyexp+"+[ ]+"

ENDIF
    i=i+1
  ELSE
    ? "That variable does not exist.  Try again."
  ENDIF
ENDDO

nomvar=SPAC(10)
@ 8+i, 0 SAY "Sum which field? " GET nomvar
```

Table 5. Continued.

```
READ
IF len(trim(nomvar))=0
  RETURN
ENDIF

SELE 2
LOCATE FOR field_name=upper(nomvar)
IF FOUND()
  IF field_type <> 'N'
    ? "That variable is not numeric.  Try again."
    RETURN
  ENDIF
ELSE
  ? "That variable does not exist.  Try again."
  RETURN
ENDIF

* create summary file
SET SAFE OFF
SELE 2
PACK
APPEND BLANK
REPL field_name WITH "TOTAL_", field_type WITH "N",;
     field_len WITH 10, field_dec WITH 0
IF (AT(".DBF",origfile)>0)
    origfile=SUBSTR(origfile,1,AT(".DBF",origfile)-1)
ENDIF
lastat=0
flname=origfile
DO WHILE (AT("\",flname)>0)
    lastat=lastat+AT("\",flname)
    flname=SUBSTR(flname,AT("\",flname)+1)
ENDDO
sumname=SUBSTR(TRIM(origfile),1,lastat+7)+"X.DBF"
CREATE &sumname FROM tempsum

USE &sumname
SELE 3

USE tempsum
GO TOP
* scanning the data file
? "Standby . . . processing"
SELE 1
INDEX ON &keyexp TO temp
SET INDEX TO temp
SET SAFE ON
keyref=&keyexp
DO WHILE .NOT. EOF()

   SELE 2
   APPEND BLANK
   SELE 3
   GO TOP
```

Table 5. Continued.

```
  DO WHILE .NOT. field_name="TOTAL_"
    SELE 2
    vari=tempsum->field_name
    REPL &vari with origfile->&vari
    SELE 3
    SKIP
  ENDDO
  SELE 1
  SUM &nomvar WHILE &keyexp=keyref TO ncount
  SELE 2
  REPL total_ WITH ncount
  SKIP
  SELE 1
  keyref=&keyexp
ENDDO

CLEAR
CLOSE DATA
ERAS temp.ndx
ERAS tempsum.dbf
USE &sumname
SET TALK ON
RETU
```

NUMERIC-TO-TEXT REPORT GENERATION

The numeric-to-text report generation format that we have established allows for numeric codes to be converted into words without altering the actual database file. This task is accomplished through the use of a linking program (LINKFILE.PRG) and support files that link numeric codes with the appropriate text for the generation of a table or a new file. By not altering the original database, further data manipulation can be accomplished based on the numeric groupings established in the coding form. The format that is presented here is a format that we have used previously, but should not be considered as the only report option. We stress changing the report form to suit presentation needs.

To use the numeric-to-text report generation program, three new database files and three new index files must first be created. The three database files should be named TAXON.DBF, ELEMENT.DBF, and PORTION.DBF. The TAXON.DBF file consists of three fields while the other two data files consist of two fields. These fields contain the numeric codes and text conversion. In the TAXON database, the first field should be named TXN, the second field should be named LABEL, and the third field COMNAME. For the ELEMENT and PORTION databases, the first field name should be EL and PE respectively and the second field should be called LABEL. The first field of these database files should be filled with the numeric codes taken from the appropriate section from the coding form. The LABEL fields should be filled with the text that corresponds to the codes from the first field. The label fields require 30 spaces for text for the TAXON and PORTION files and 27 spaces for the ELEMENT file. The COMNAME field of the TAXON file will also require 30 spaces. After these database files have been created and coding form data input into the proper fields, these files must be indexed. These files must be indexed on the first field that contains the numeric codes. The indexed files should be named TXN.NDX, EL.NDX, and PE.NDX, respectively. These files contain the correlations for the LINKFILE program (Table 6). As an example, number 150 in EL.NDX produces the word "Clavicle" when the report is generated.

Next, a dBase report form must be created. Follow the normal dBase procedures for creating a report form named LISTFILE. Before a report form can be created that will work with LINKFILE, LINKFILE must first be executed (DO LINKFILE) and will ask the user for the file that is to be converted. Once the file is converted, "FXXN" will be displayed on the command line. This step is necessary so that information from fields two through four (Table 7) will be accepted by the report form. These fields are responsible for generating text from the numeric database.

When creating the report form, under the "Options" column, the subheading "Page width" must be set to equal or exceed the width of all of the columns for the report. We use 177 spaces for "Page Width." Under "Column," "Contents" is the name of the field from which data from the dBase file will be obtained. Listed under "Contents," is "Heading." "Heading" can be used to give a name to the column in the report form. Next is "Width." "Width" is the number of spaces needed for the column once it has been converted into words. The next two subheadings are "Decimal places" and "Total this column." Ignore the former and answer "NO" to "Total this column." Table 7 provides a listing of what information should be placed into the subheadings under "Column." Nine field descriptions are listed in this table to aid in report form generation.

NOTE: FOR THE LINKFILE PROGRAM TO WORK, THE LINKFILE PROGRAMS AND SUPPORTING FILES MUST BE LOCATED IN THE DEFAULT DRIVE AND DIRECTORY. THE DATA TO BE CONVERTED MAY BE LOCATED IN EITHER THE DEFAULT DRIVE OR ANOTHER DRIVE OR DIRECTORY.

Once each of these files has been set up along with the report form, and data has been entered into a database, a report can then be generated. In order to generate a report, the first step is to use the LINKFILE program. This is accomplished by typing DO LINKFILE. LINKFILE will then prompt the user for the name of the file to be converted. Here, the user types in the name of the dBase file that contains the zooarchaeological data. After LINKFILE has been executed, "FXXN" will be displayed on the command line. At the dot prompt, type, REPORT FORM LISTFILE TO PRINT. Instead of PRINT, FILE may be substituted. PRINT sends the converted report to the printer. If FILE is chosen, the converted information is saved to an ASCII file. Before an ASCII file will be created, dBase will prompt the user for a file name for the ASCII file. This ASCII file can then be interfaced with word processing, statistical, and graphics programs. Both will list the converted data on the screen. If at a later time new tables are to be generated, the above procedures must be repeated.

The end product of this report form will look like the information provided in Table 8. This table was created from the information from Table 1. Several categories were omitted in this table because they were not used in the analysis. For the categories "Weathering," "Breakage," "Burning," and "Cut marks," numeric codes have been retained in order to save space. A key has been provided explaining the codes for each. We have chosen this format for these last four categories because it is efficient in terms of space. We have displayed it here as an example of an alternative way of presenting the data.

The LINKFILE and "report form" (dBase procedure) are a quick and reliable way to convert coded data back into text. Major advantages of using this program include quick table generation and continuity in description and spelling. In this way, organized and succinct tables can be generated with minimal effort. Also, LINKFILE can be used in conjunction with the XTOT program. In this way, NISP and MNI estimations can be more easily determined.

Table 6. LINKFILE Program.

```
*LINKFILE.PRG

CLEAR
fname = SPACE(25)
@10,10 SAY "What is the name of the file that you want converted?"
@11,25 GET fname
READ
fname=TRIM(UPPER(fname))
USE &fname ALIAS FXXN
SELECT 2
USE TAXON INDEX TXN
SELECT 3
USE ELEMENT INDEX EL
SELECT 4
USE PORTION INDEX PE
SELECT 1
SET RELATION TO TXN INTO TAXON
SELECT 2
relat="FXXN->EL"
SET RELATION TO &relat INTO ELEMENT
SELECT 3
relat="FXXN->PE"
SET RELATION TO &relat INTO PORTION
SELECT 1
```

Table 7. Report Form Field Information.

Field	Subheading	Information
1	Contents Heading Width	Qty QTY 5
2	Contents Heading Width	TAXON->LABEL Taxon 30
3	Contents Heading Width	ELEMENT->LABEL Element 27
4	Contents Heading Width	PORTION->LABEL Portion 30
5	Contents Heading Width	IIF(SX=0,"",IIF(SX=1,"Male",IIF(SX=2,"Female","ERROR"))) Sex 6
6	Contents Heading Width	IIF(SD=0,"",IIF(SD=1,"Left",IIF(SD=2,"Right",IIF(SD=3,"Axial","ERROR")))) Side 5
7	Contents Heading Width	IIF(AG=0,"",IIF(AG=3,"Fetal/Infant",IIF(AG=5,"Subadult",IIF(AG=6,"Adult",IIF(AG=8,"Old adult","ERROR"))))) Age 9
8	Contents Heading Width	STR(W,1)+STR(BK,1)+STR(B,1)+STR(G,1)+STR(CT,1) WBBGC; K 5
9	Contents Heading Width	Comments Comments 30

Table 8. Data From Table 1 Converted into a Report.

Room	Qty	Taxon	Element	Portion of Element	SIDE	Age	WBBCK	Comments
2	1	Sylvilagus sp.	Pelvis	Acetabulum w/il., isch., pub.	Right		1100	
2	1	Sylvilagus sp.	Permanent tooth	Lower PM3	Left		1000	
2	1	Spermophilus sp.	Mandible	Complete or nearly complete	Left	Subadult	1000	
2	2	Spermophilus sp.	Permanent tooth	Lower PM4	Left	Subadult	1000	
5	2	Mammalia	Rib	Shaft fragment			1100	
5	1	Lepus sp.	Calcaneus	Complete or nearly complete	Left		1000	
5	1	Geomyidae	Pelvis	Os coxae	Left		1000	
5	1	Geomyidae	Mandible	Horizontal ramus portion	Left		1100	
5	10	Mammalia	Indeterminate	Indeterminate			1100	
5	1	Geomyidae	Permanent tooth	Lower I1	Right		1000	
5	1	Sciuridae	Scapula	Glenoid fossa & incom. blade	Right		1100	
5	1	Lepus sp.	Tibia	Proximal end	Right		1100	
5	1	Aves	Radius	Complete or nearly complete		Subadult	1000	
5	1	Zenaida sp.	Carpometacarpus	Complete or nearly complete	Left		1000	
5	1	Geomyidae	Cranium	Premaxilla	Left		1100	
5	1	Geomyidae	Permanent tooth	Upper PM4	Right		1000	
5	1	Cricetidae	Tibia	Proximal end	Right		1100	
5	2	Leporidae	Lumbar vertebra	Complete or nearly complete	Axial		1000	
5	1	Sylvilagus sp.	Femur	Distal end			1100	
5	1	Neotoma sp.	Permanent tooth	Lower M1	Right		1000	
5	1	Spermophilus variegatus	Permanent tooth	Lower I1	Left		1100	
8	1	Sylvilagus sp.	Femur	Complete shaft	Right	Subadult	1100	
8	1	Lepus sp.	Humerus	Distal portion of shaft	Left		1110	
8	1	Sciuridae	Mandible	Horizontal ramus portion	Right		1100	
8	1	Sylvilagus sp.	Radius	Complete or nearly complete	Right		1000	
8	1	Artiodactyla	Lumbar vertebra	Complete minus spinous process	Axial		1100	
9	1	Odocoileus sp.	Calcaneus	Complete or nearly complete	Right		1001	

Key:
W - Weathering, 0=Absent, 1=Slight, 2=Marked
Bk - Breakage, 0=Absent, 1=Angular, 2=Spiral, 3=Both
B - Burning, 0=Absent, 1=Burned
C - Cut marks, 0=Absent, 1=Present

CONDENSE FILE PROCEDURES

The following procedures (Table 9) reduce a dBase file down to the fewest number of unique lines. The command phrases are in all capital letters and represent the actual commands. Words in parentheses in Table 9 are either descriptions of the preceding commands or are secondary commands that are performed by the user. These procedures can be advantageous when large data sets are to be manipulated and published. This will combine all lines of data that are exactly the same into one line, disregarding quantity information, and will then sum the quantity field. In this way, all of the same elements, broken in the same way, from the same provenience, etc., will be combined and totaled. However, for this program to work properly, the quantity field MUST be the first field listed and must be named "QTY." If an error message of "Data has been lost" appears after the program has run, the quantity field was too small to handle the total of a recombined line(s). That is, the total exceeded 99,999 specimens. If this is the case, modify the backup of your original data set and increase the spaces in the quantity field from five to six. Then, follow the above procedures again.

Table 9. dBase File Condensing Steps.

1. USE _____.DBF [R] (this is the dBase file that is to be reduced)
1a. **The QUANTITY field must be the first field. If it is not, then modify the structure of the file appropriately.**
2. COPY STRUCTURE TO _____.DBF [R] (This file will be the finished product file in which all of the records have been collapsed)
3. COPY TO TEMP1 SDF [R] (no period before SDF)
4. COPY TO TEMPSTR STRUCTURE EXTENDED [R]
5. USE TEMPSTR [R]
6. SUM [R] (Note the total of the sum for FIELD LEN & mentally subtract the figure for the QTY field)
7. DELETE FOR RECNO() > 2 [R]
8. PACK [R]
9. BROWSE [R]
9a. (Now rename the second field listed to STRING)
9b. (Change second field type to "CHARACTER")
9c. (Change second field length to equal FIELD LEN minus the QTY field length)
9d. CONTROL-END
10. CREATE TEMPFIL2 FROM TEMPSTR [R]
11. APPEND FROM TEMP1 SDF [R] (append from the SDF file created in step 3)
12. INDEX ON STRING TO TEMPSTR [R]
13. TOTAL ON STRING TO TEMPTOT [R] (This is the file reduced, but the proper fields must now be re-established)
14. USE TEMPTOT [R]
15. COPY TO TEMPEND SDF [R]
16. USE _____.DBF [R] (This is the file designated from step 2 to be the end product file)
17. APPEND FROM TEMPEND SDF [R]

18. RUN DEL TEMP*.* [R] (This erases the extra files created during this process)

[R] indicates that the RETURN or ENTER key should be pressed

DATABASE LINK PROCEDURES

The following commands are used to link a supplemental data file with a main data file. A supplemental data file may be any additional file that contains data that the user wishes to add to the main data file. For example, we normally record all faunal data by either lot or field sack number in the main data file. In order to add all of the provenience data connected to each lot number, a separate file with the lot number and provenience data must first be created. This is the supplemental data file. Also, provenience information contained in the supplemental file must have corresponding fields in the main data file. If the fields do not match exactly, then the data will not be correctly transferred. All of the words in capitals are commands that must be used (Table 10). Once the supplemental data file and additional fields have been created, the next step is the transferring of the data.

The command phrases are in capital letters. Words that are not in capital letters represent information that must be provided by the researcher. Any punctuation presented in the command phrase must be used. Words in parentheses are explanations for the commands that precede them. Before beginning this set of commands, be sure that a backup of the original data file has been made.

Table 10. Main and Supplemental Data
File Linking Procedures.

1. CLEAR ALL (Clears all files that may be in use)

2. USE supplemental filename ("supplement filename" is the file that provides the additional data for the main file)

3. INDEX ON fieldname TO filendx (Here, fieldname is the field that is contained in both the data file and the main file. This is the field upon which the linking will occur and contains the field sack or lot numbers or such. "filendx" is the name of the index file that will be created. This file can be given any name. It is important to note that the field on which the link depends must be of the same name (exactly), same length, and same field type in both files. Otherwise, the procedure will not work)

4. SELECT 2 (this command allows you to work with more than one file in dBase at the same time)

5. USE main filename ("main filename" is the main data file into which you wish to place data from the supplemental data file)

6. SET RELATION TO fieldname INTO supplemental filename (this establishes the communication between the two files)

7. REPLACE ALL north WITH supplemental filename->north, west WITH supplemental filename->west, lvl WITH supplemental filename->lvl (The fieldnames "north," "west," and "lvl" are all examples. You must insert the field names that you wish the data in the supplemental file to replace in the main file. The "->" must be used for this to work. In order to have multiple fields replaced, the different instructions must only be separated at the appropriate points by a comma--see the above example)

8. Now verify that the appropriate fields have been filled with data. If a mistake is made in the process, all of the functions will run, but no data will be replaced.

9. Now erase the "filendx" file that has been created. This file should be deleted because it will no longer be needed.

REFERENCES CITED

Aaris-Sorensen, Kim
1981   A Classification Code and Computerized Data-Analysis for Faunal Materials from Archaeological Sites. *Ossa* 8:3-29.

Armitage, P. L.
1978   A System for the Recording and Processing of Data Relating to Animal Remains from Archaeological Sites. In *Research Problems in Zooarchaeology: Occasional Publication No. 3*, edited by D. R. Brothwell, K. D. Thomas, and Juliet Clutton-Brock, pp. 39-45. Institute of Archaeology, London.

Ashton-Tate
1986   dBase III Plus Database Program. Torrance, California.

1989   dBase IV Database Program. Torrance, California.

Bailey, Louis, Kenneth Steigman, and Barbara Peterman
1990   Application of dBase III Plus to Database Needs of Small Museums. *Curator* 33:207-216.

Baker, Barry W., Brian S. Shaffer, Kristin D. Sobolik, and
D. Gentry Steele
1991   Chapter 7. Faunal Analysis, Part I: Analysis of the Vertebrate Faunal Remains. In *Alabonson Road: Early Ceramic Adaptation to the Inland Coastal Prairie Zone, Harris County, Southeast Texas*. Edited by H. Blaine Ensor and David L. Carlson, pp. 139-161, Reports of Investigations No. 8. Archeological Research Laboratory, Texas A&M University, College Station.

Baker, J. and D. Brothwell
1980   *Animal Diseases in Archaeology*. Academic Press, New York.

Bayham, Frank E.
1982   *A Diachronic Analysis of Prehistoric Animal Exploitation at Ventana Cave*. Ph.D. dissertation, Arizona State University, Tempe.

Behrensmeyer, Anna K.
1978   Taphonomic and Ecologic Information from Bone Weathering. *Paleobiology* 4:150-162.

Binford, Lewis R.
1981   *Bones: Ancient Men and Modern Myths*. Academic Press, New York.

Bonnichsen, Robson and David Sanger
1977   Integrating Faunal Analysis. *Canadian Journal of Archaeology* 1:109-133.

Brumley, John H.
1973   Quantitative Methods in the Analysis of Butchered Faunal Remains: A Suggested Approach. *Archaeology in Montana* 12:1-40.

Bryan, Kelly, Tony Gallucci, and David H. Riskind
  1984   *A Checklist of Texas Birds*, Technical Series No. 32. Texas Parks & Wildlife Department, Austin.

Burt, William H. and Richard P. Grossenheider
  1976   *A Field Guide to the Mammals of America North of Mexico*. Houghton Mifflin Company, Boston.

Campana, Douglas V. and Pam J. Crabtree
  1987   Animals- A C Language Computer Program for the Analysis of Faunal Remains and Its Use in the Study of Early Iron Age Fauna from Dun Ailinne. *ArchaeoZoologia* 1(1):57-68.

  1988   "Book Reviews." *Zooarchaeological Research News* 7(1):13-14.

Cannon, Debbie Yee
  1987   *Marine Fish Osteology*. Archaeology Press, Simon Fraser University, Burnaby, B.C.

Casteel, Richard W.
  1976   *Fish Remains in Archaeology*. Academic Press, New York.

Clutton-Brock, J.
  1975   A System for the Retrieval of Data Relating to Animal Remains from Archaeological Sites. In *Archaeozoological Studies*, edited A. T. Clason, pp. 21-34. American Elsevier Publishing Company, New York.

Conant, Roger
  1975   *A Field Guide to Reptiles and Amphibians of Eastern and Central North America*, 2nd edition. Houghton Mifflin Company, Boston.

Courtemanche, Michelle and Vianney Legendre
  1985   *Os de Poissons: Nomenclature Codifiee Noms Francais et Anglais*. Osteotheque de Montreal, Inc., Montreal, Quebec.

Cruz-Uribe, Kathryn and Richard G. Klein
  1986   Pascal Programs for Computing Taxonomic Abundance in Samples of Fossil Mammals. *Journal of Archaeological Science* 13:171-187.

Desse, Jean and Louis Chaix
  1986   An Operational Osteometric Database. In *Database Management and Zooarchaeology*, edited by Louise H. van Wijngaarden-Bakker, *Pact* 14:93-98.

Desse, Jean, Louis Chaix, and Nathalie Desse-Berset
  1986   "Osteo" Base-Reseau de Donnees Osteometriques pour l' Archeozoologie. Centre National de laRecherche Scientifique, Paris.

Dixon, James R.
  1987   *Amphibians and Reptiles of Texas*. Texas A&M University Press, College Station.

Dobney, Keith and Kevin Rielly
   1988   A Method for Recording Archaeological Animal Bones: the Use of Diagnostic Zones. *Circaea* 5:79-96.

Findley, James S., Arthur H. Harris, Don E. Wilson, and Clyde Jones
   1975   *Mammals of New Mexico*. University of New Mexico Press, Albuquerque.

Gifford, Diane P. and Diana C. Crader
   1977   A Computer Coding System for Archaeological Faunal Remains. *American Antiquity* 42:225-238.

Gifford-Gonzalez, Diane and Barbara Wright
   1986   A Data Management and Table Formatting System for Vertebrate Remains. In *Database Management and Zooarchaeology*, edited by Louise H. van Wijngaarden-Bakker, *Pact* 14:137-164.

Gilchrist, R. and H. C. Mytum
   1986   Experimental Archaeology and Burnt Animal Bone from Archaeological Sites. *Circaea* 4:29-38.

Grant, Annie
   1975   The Use of Tooth Wear as a Guide to the Age of Domestic Animals - A Brief Explanation. In *Excavation at Porchester Castle I: Roman. B. Cunliffe*. London: Society of Antiquaries Research Report, No. 32.

   1978   Variation in Dental Attrition in Mammals and its Relevance to Age Estimation. In *Research Problems in Zooarchaeology*. D. R. Brothwell, K. D. Thomas, S.Clutton-Brock, eds. pp. 103-106. London Institute of Archaeology, Occasional Publications, No. 3.

   1982   The Use of Tooth Wear as a Guide to the Age of Domestic Ungulates. In *Ageing and Sexing Animal Bones from Archaeological Sites*. B. Wilson, C. Grigson, and S. Payne, eds. pp. 91-108. B.A.R., British Series 109.

Grayson, Donald K.
   1984   *Quantitative Zooarchaeology: Topics in the Analysis of Archaeological Faunas*. Academic Press, Inc.

   1988   3. Last Supper Cave. In *Danger Cave, Last Supper Cave, and Hanging Rock Shelter: The Faunas*, edited by Donald K. Grayson. Anthropological Papers of the American Museum of Natural History 66(1).

Guilday, John E., Paul W. Parmalee, and Donald P. Tanner
   1962   Aboriginal Butchering Techniques at the Eschelman Site (36 La 12), Lancaster County, Pennsylvania. *Pennsylvania Archaeologist* 32:59-83.

Hall, E. Raymond
   1981   *The Mammals of North America*. John Wiley & Sons, New York.

Harrison, Richard J. and Judith E. King
  1965   *Marine Mammals*. Hutchinson University Library, London.

Hellier, Jana and Cristi Assad
  n.d.   Coding form on file, Department of Anthropology, Texas A&M University, College Station.

Hillson, Simon
  1986   *Teeth*. Cambridge University Press, New York.

Johnson, Eileen
  1985   Current Developments in Bone Technology. In *Advances in Archaeological Method and Theory, Volume 8*, edited by Michael B. Schiffer, pp. 157-235. Academic Press, New York.

Klein, Richard G. and Kathryn Cruz-Uribe
  1984   *The Analysis of Animal Bones from Archaeological Sites*. The University of Chicago Press, Chicago.

Koster, William J.
  1957   *Guide to the FISHES of New Mexico* (2nd ed.). University of New Mexico Press, Albuquerque.

Levine, Marsha A.
  1982   The Use of Crown Height Measurements and Eruption Wear Sequences to Age Horse Teeth. In *Ageing and Sexing Animal Bones from Archaeological Sites*. B. Wilson, C. Grigson, and S. Payne, eds. pp. 223-254. B.A.R., British Series 109.

Liem, F. Karel
  1963   *The Comparative Osteology and Phylogeny of the Anabantoidei (Teleostei, Pisces)*. University of Illinois Press, Urbana.

Limp, W. Fedrick, James A. Farley, and Carol Andrews
  1986   A Userfriendly Menudriven Microcomputer Based System for Artifact Inventory and Analysis. In *Database Management and Zooarchaeology*, edited by Louise H. van Wijngaarden-Bakker, *Pact* 14:73-86.

Lyman, R. Lee
  1977   Analysis of Historic Faunal Remains. *Historical Archaeology* 11:67-73.

  1991   *Prehistory of the Oregon Coast: the Effects of Excavation Strategies and Assemblage Size on Archaeological Inquiry*. Academic Press, San Diego.

Lyman, R. Lee and Gregory L. Fox
  1989   A Critical Evaluation of Bone Weathering as an Indication of Bone Assemblage Formation. *Journal of Archaeological Science* 16:293-317.

McArdle, John
  1975-   A Numerical (Computerized) Method for Quantifying
  1977   Zooarchaeological Comparison. *Paleorient* 3:181-190.

McClane, A. J.
1974 *Field Guide to Saltwater Fishes of North America*. Holt, Rinehart and Winston, New York.

Meadow, Richard H.
1978 "BONECODE" - A System of Numerical Coding for Faunal Data from Middle Eastern Sites. In *Approaches to Faunal Analysis in the Middle East*, edited by Richard H. Meadow and Melinda A. Zeder, pp. 169-186. Bulletin of the Peabody Museum of Archaeology and Ethnology, Cambridge.

Muniz, Arturo Morales
1988 On the Use of Butchering as a Paleocultural Index: Proposal of a New Methodology for the Study of Bone Fracture from Archaeological Sites. *ArchaeoZoologia* 2:111-150.

Munzel, S.
1986 Coding System for Bone Fragments. In *Database Management and Zooarchaeology*, edited by Louise H. van Wijngaarden-Bakker, *Pact* 14:193-195.

Munzel, Susanne C.
1988 Quantitative Analysis and Archaeological Site Interpretation. *ArchaeoZoologia* 2:93-110.

Nelson, Joseph S.
1984 *Fishes of the World*. John Wiley & Sons, New York.

Nichol, R. K. and G. A. Creak
1979 Matching Paired Elements Among Archaeological Bone Remains: A Complete Procedure and Some Practical Limitations. *Newsletter of Computer Archaeology* 14:6-16.

Olsen, Stanley J.
1968 *Fish, Amphibian and Reptile Remains from Archaeological Sites*. Papers of the Peabody Museum of Archaeology and Ethnology, Harvard University Vol. 56.

Parker, Sandra and Michael Kaczor
1986 The DELOS Archaeological Database Management System. In *Database Management and Zooarchaeology*, edited by Louise H. van Wijngaarden-Bakker, *Pact* 14:45-71.

Payne, S.
1973 Kill-off Patterns in Sheep and Goats: the Mandibles from Asvan Kale. *Anatolian Studies* 23:281-303.

Peters, Joris
1987 Cuboscaphoids, Naviculo-cuboids, Language Barriers and the Use of Standardised Osteological Nomenclatures in Archaeozoological Studies. *ArchaeoZoologia* 1(2):43-46.

Powell, Joseph F.
   1991   6. Human Skeletal Remains from Skyline Shelter (41VV930), Val Verde County, Texas. In *Papers on Lower Pecos Prehistory*, edited by Solveig A. Turpin, pp. 149-173, Studies in Archeology No. 8. Texas Archeological Research Laboratory, University of Texas, Austin.

Redding, R. W., J. Wheeler Pires-Ferreira, and M. Zeder
   1975-   A Proposed System for Computer Analysis of
   1977   Identifiable Faunal Material from Archaeological Sites. *Paleorient* Vol 3.

Redding, Richard W., Melinda A. Zeder, and John McArdle
   1978   "BONESORT II" - A System for the Computer Processing of Identifiable Faunal Material. In *Approaches to Faunal Analysis in the Middle East*, edited by Richard H. Meadow and Melinda A. Zeder, pp. 135-147. Bulletin of the Peabody Museum of Archaeology and Ethnology, Cambridge.

Reed, C. A.
   1971   New Method for Recording and Analyzing Faunal Material from Archaeological Sites. Paper circulated at the Section on Animal Domestication of the Third International Congress of the Museums of Agriculture, (Budapest, 19-23 April 1971).

Robbins, Chandler S., Bertel Bruun, and Herbert S. Zim
   1983   *A Guide to Field Identification Birds of North America*. Golden Press, New York.

Rojo, Alfonso I.
   1991   *Dictionary of Evolutionary Fish Osteology*. CRC Press, Ann Arbor.

Romer, Alfred Sherwood and Thomas S. Parsons
   1986   *The Vertebrate Body*, 6th ed. Saunders College Publishing, New York.

Sellards, E. H.
   1952   *Early Man in America*. Greenwood Press, New York.

Shaffer, Brian S.
   1989   Appendix I. Analysis of the Faunal Remains Recovered from 41GM23. In *Archeological and Geomorphological Investigations of 41GM23, Rocky Creek, Grimes County, Texas*, by H. Blaine Ensor, Randy Korgel, Saul Aronow, and C. S. Mueller-Wille, pp. 61-67. Reports of Investigations No. 10. Archeological Research Laboratory, Texas A&M University, College Station.

   1990   Appendix III. Analysis of the Faunal Remains. In *Archeological Investigations at Fort Brown (41CF96) Cameron County, Texas*, by Shawn B. Carlson, Joe Saunders, Frank Winchell, and Bruce Aiken, pp. 135-148, Reports of Investigations No. 11. Archeological Research Laboratory, Texas A&M University, College Station.

1991   *The Economic Importance of Vertebrate Faunal Remains from the NAN Ruin (LA15049), a Classic Mimbres Pueblo Site, Grant County, New Mexico*. M.A. Thesis, Department of Anthropology, Texas A&M University, College Station.

Shaffer, Brian S. and Barry W. Baker
1988   Appendix I. Analysis of the Vertebrate Faunal Remains Recovered at 41HR541. In *Archeological Investigations at a Late Ceramic Period Bison Kill Site, (41HR541), Whiteoak Bayou, Harris County, Texas*, by Mary Jane McReynolds, Randy Korgel, and H. Blaine Ensor, pp. 73-91, Reports of Investigations No. 7. Archeological Research Laboratory, Texas A&M University, College Station.

1990   Faunal Analysis. *Anthropology Newsletter*, Soft.Where review column, p. 2, March.

Shaffer, Brian S. and Elizabeth Miller
n.d.   Analysis of the Vertebrate Fauna Recovered from the NAN-15 Ruin, A Late Classic Mimbres Pueblo. Report on file, Department of Anthropology, Texas A&M University, College Station.

Shipman, Pat, Giraud Foster, and Margaret Schoeninger
1984   Burnt Bones and Teeth: an Experimental Study of Color, Morphology, Crystal Structure and Shrinkage. *Journal of Archaeological Science* 11:307-325.

Shively, M. J.
1984   *Veterinary Anatomy: Basic, Comparative, and Clinical*. Texas A&M University Press, College Station, Texas.

Sobolik, Kristin D. and D. Gentry Steele
n.d.   An Atlas of Turtles to Facilitate Archaeological Identification. Manuscript on file, Department of Anthropology, Texas A&M University, College Station.

Spennemann, Dirk H. R. and Sarah M. Colley
1989   Fire in a Pit: the Effects of Burning on Faunal Remains. *ArchaeoZoologia* 3:51-64.

Stebbins, Robert C.
1985   *A Field Guide to Western Reptiles and Amphibians*. Houghton Mifflin Company, Boston.

Storer, David Humphrey
1972   *A Synopsis of the Fishes of North America*. A. Asher & Co.B.V., Amsterdam.

Tamplin, Morgan, Shawn Haley, and Deborah DeHetre
1983   Small Mammal Butchering in Prehistory: Beaver and Muskrat Remains from the Pas Reserve Site, Manitoba. *Manitoba Archaeological Quarterly* 7:5-33.

Uerpmann, Hans-Peter
   1978   The "KNOCOD" System for Processing Data on Animal Bones from Archaeological Sites. In *Approaches to Faunal Analysis in the Middle East*, edited by Richard H. Meadow and Melinda A. Zeder, pp. 149-167. Bulletin of the Peabody Museum of Archaeology and Ethnology, Cambridge.

Van Tyne, Josselyn and Andrew J. Berger
   1976   *Fundamentals of Ornithology*, 2nd ed. John Wiley & Sons, New York.

van Wijngaarden-Bakker, Louise H.
   1986   Scientific Information Retrieval of Zooarchaeological Data. In *Database Management and Zooarchaeology*, edited by Louise H. van Wijngaarden-Bakker, *Pact* 14:115-128.

von den Driesch, Angela and Joachim Boessneck
   1975   Schmittspuren en neolithischen Tierknochen. *Germania* 53:1-23.

Wheeler, Alwyne and Andrew K. G. Jones
   1989   *Fishes*. Cambridge University Press, New York.

Whitaker, John O., Jr., and Robert Elman
   1987   *The Audubon Society Field Guide to North American Mammals*. Alfred A. Knopf, Inc., New York.

White, T. E.
   1953   A Method of Calculating the Dietary Percentages of Various Food Animals Utilized by Aboriginal Peoples. *American Antiquity* 18:396-398.

Yates, Bonnie C.
   n.d.   Coding form on file, University of North Texas, Denton.

Zimmerman, Laurie S.
   1990   Appendix E: Vertebrate Fauna. In *Hunter-Fisher-Gatherers on the Upper Texas Coast: Archeological Investigations at the Peggy Lake Disposal Area, Harris County, Texas*, by Eloise F. Gadus and Margaret Ann Howard, pp. 335-355, Reports of Investigations No. 74. Prewitt and Associates, Inc., Austin, Texas.

UMMA Backlist

Four series of publications are available from the Publications Office of the University of Michigan Museum of Anthropology. The Occasional Contributions, published from 1932 through 1956, and the Anthropological Papers, begun in 1949, are two series of short monographs, while the Memoirs, first published in 1970, are longer, more detailed studies. The fourth series, Technical Reports, begun in 1971, are brief, highly technical discussions of recent advances in several areas of anthropological study. New subseries will be added to the Technical Reports from time to time. Contributions to all of the series are prepared by staff members, associates, and friends of the Museum and include descriptions of museum collections and field work, results of research in various anthropological fields, and discussions of field and museum techniques.

Henry T. Wright, Director
Museum of Anthropology

The books below may be ordered from the Museum of Anthropology, 4009 Museums, University of Michigan, Ann Arbor, MI 48109. Libraries and members of the Michigan Archaeological Society receive a 20% discount. Checks must be in U.S. funds drawn on a U.S. bank. Please include $2 postage for all orders less than $10, $4 postage for orders over $10. Prepayment is required.

Occasional Contributions

6. The Younge Site: An Archaeological Record from Michigan, by Emerson F. Greenman. 1937. Reprinted 1967. Pages 172, 33 plates, 9 figures, 10 maps. Price $3.
15. Araucanian Culture in Transition, by Mischa Titiev. 1951. Pages 164, 17 plates, 9 figures, 2 maps. Price $2.50.

Anthropological Papers

13. The Puerto Rican Population: A Study of Human Biology, by Frederick P. Thieme. 1959. Pages 156, 4 figures, 2 maps. Price $2.50.
14. Tell Toqaan: A Syrian Village, by Louise E. Sweet. 1960. Pages 280, 54 figures. Price $2.50.
39. Rules of Descent: Studies in the Sociology of Parentage, by Guy E. Swanson. 1969. Pages 108. 4 figures, 7 tables. Price $2.
41. The Archaeology of Summer Island: Changing Settlement Systems in Northern Lake Michigan, by David S. Brose. 1970. Pages 236, 31 tables, 17 figures, 35 plates. Price $3.
42. The Occupations of Migrants in Ghana, by Polly Hill. 1970. Pages 84, 11 tables. Price $2.
44. Property Control and Social Strategies: Settlers on a Middle Eastern Plain, by Barbara C. Aswad. 1971. Pages 180, 16 tables, 33 figures, 2 plates. Price $3.50.
48. The Wardell Buffalo Trap 48 SU 301: Communal Procurement in the Upper Green River Basin, Wyoming, by George C. Frison. 1973. Pages 111, 29 figures, 6 tables, 14 plates. Price $3.
50. Faction and Conversion in a Plural Society: Religious Alignments in the Hindu Kush, by Robert Leroy Canfield. 1973. Pages 142, 11 figures, 4 tables, 1 appendix. Pricce $3.
55. The Ait Ndhir of Morocco: A Study of the Social Transformation of a Berber Tribe, by Amal Rassam Vinogradov. 1974. Pages 128, 11 figures, 13 plates. Price $4.
59. An Analysis of Effigy Mound Complexes in Wisconsin, by William M. Hurley. 1975. Pages 466, 63 figures, 48 tables, 45 plates. Price $8.
62. The Demography of the Semai Senoi, by Alan G. Fix. 1977. Pages 123, 17 figures, 38 tables. Price $5.
63. Economic and Social Organization of a Complex Chiefdom: The Halelea District, Kauai, Hawaii, by Timothy Earle. 1978. Pages 205, 27 figure, 7 tables, 6 plates. Price $6.
64. Wāsita in a Lebanese Context: Social Exchange Among Villagers and Outsiders, by Frederick Charles Huxley. 1978. Pages 174, 6 figures, 47 tables, 5 plates. Price $6.
65. Meadowood Phase Settlement Patterns in the Niagara Frontier Region of Western New York State, by Joseph E. Granger, Jr. 1978. Pages 403, 73 figures, 113 tables, 35 plates. Price $8.
66. The Snodgrass Site of the Powers Phase of Southeast Missouri, by James E. Price and James B. Griffin. 1979. Pages 189, 80 figures, 2 tables, 17 plates. Price $6.

67. The Nature and Status of Ethnobotany, edited by Richard I. Ford. 1978. Pages 428, 33 figures, 28 tables, 24 plates. Price $10.
68. The Biological and Social Analyses of a Mississippian Cemetery from Southeast Missouri: The Turner Site, 23BU21A, by Thomas K. Black III. 1979. Pages 170, 7 figures, 69 tables, 10 plates. Price $6.
69. The Ait Ayash of the High Molouuya Plain: Rural Social Organization in Moroco, by John Chiapuris. 1980. Pages 186, 15 figures 9 maps, 12 plates. Price $6.
70. An Early Woodland Community at the Schultz Site 20SA2 in the Saginaw Valley and the Nature of Early Woodland Adaptation in the Great Lakes Region, by Doreen Ozker. 1982. Pages 273, 27 tables, 33 figures, 15 plates. Price $10.
71. Persian Diary, 1939–1941, by Walter N. Koelz. 1983. Pages 227, 2 maps, 68 photos. Price $10.
72. Lulu Linear Punctated: Essays in Honor of George Irving Quimby, edited by Robert C. Dunnell and Donald K. Grayson. 1983. Pages 354, 39 figures, 19 plates, 20 tables. Price $12.
73. Paleoethnobotany of the Kameda Peninsula Jomon, by Gary W. Crawford. 1983. Pages 200, 27 figures, 12 tables, 23 plates. Price 48.
74. The Archaeology of the Sierra Blanca Region of Southeastern New Mexico, by Jane Holden Kelley. 1983. Pages 527, 85 figures, 10 maps, 41 tables, 87 plates. Price $15.
75. Prehistoric Food Production in North America, edited by Richard I. Ford. 1985. Pages 411, 39 figures, 22 tables. Price $15.
76. Primitive Polluters: Semang Impact on the Malaysian Tropical Rain Forest Ecosystem, by A. Terry Rambo. 1985. Pages 104, 5 figures, 5 tables, 16 plates. Price $8.
77. Jumano and Patarabueye: Relations at La Junta de los Rios, by J. Charles Kelley. 1986. Pages 180, 14 figures, 9 plates. Price $10.
78. Protohistoric Yamato: Archaeology of the First Japanese State, by Gina L. Barnes. 1988. Pages 473, 94 figures, 17 tables. Price $15.
79. The Foxie Otter Site: A Multicomponent Occupation North of Lake Huron, by Christopher C. Hanks. 1988. Pages 203, 21 figures, 14 tables. Price $12.
80. Living in a Lean-to: Philippine Negrito Foragers in Transition, by Navin K. Rai. 1990. Pages 184, 4 figures, 10 plates, 12 appendices. Price $12.
81. The Bridgeport Township Site: Archaeological Investigation at 20SA620, Saginaw County, Michigan, edited by John O'Shea and Michael Shott. 1990. Pages 326, 50 figures, 68 tables, 1 appendix. Price $15.00
82. Maguey Utilization in Highland Central Mexico: An Archaeological Ethnography, by Jeffrey R. Parsons and Mary H. Parsons. 1990. Pages 388, 46 figures, 172 plates, 26 tables, appendix. Price $22.
83. Vertebrate Faunal Remains from Grasshopper Pueblo, Arizona, by John W. Olsen. 1990. Pages 200, 17 figures, 6 tables, appendix. Price $15.
84. Debating Oaxaca Archaeology, edited by Joyce Marcus. 1990. Pages 270, 79 illustrations, 11 tables. Price $18.
85. Profiles in Cultural Evolution: Papers from a Conference in Honor of Elman R. Service, edited by A. Terry Rambo and Kathleen Gillogly. 1991. Pages 450, 33 figures, 11 tables. Price $20.

Memoirs

2. The Burial Complexes of the Knight and Norton Mounds in Illinois and Michigan, by James B. Griffin, Richard E. Flanders and Paul F. Titterington. 1970. Pages 216, 177 plates. Price $7.
3. Prehistoric Settlement Patterns in the Texcoco Region, Mexico, by Jeffrey R. Parsons. 1971. Pages 447, 8 tables, 14 maps, 88 figures, 57 plates. Price $8.
4. The Schultz Site at Green Point: A Stratified Occupation Area in the Saginaw Valley of Michigan, edited by James E. Fitting. 1972. Pages 317, 84 figures, 70 tables, 2 appendixes. Price $8.
7. Formative Mesoamerican Exchange Networks with Special Reference to the Valley of Oaxaca, by Jane W. Pires-Ferreira. Prehistory and Human Ecology of the Valley of Oaxaca, Vol. 3. Pages 111, 44 figures, 27 tables, 4 plates. Price $6.
8. Fábrica San José and Middle Formative Society in the Valley of Oaxaca, by Robert D. Drennan. Prehistory and Human Ecology of the Valley of Oaxaca, Vol. 4. 1975. Pages 300, 10 tables, 2 maps, 75 figures, 29 plates. Price $8.
9. Studies in the Archaeological History of the Deh Luran Plain, by Frank Hole. 1977. Pages 369, 94 tables, 119 illustrations, 55 plates. $10.
10. Part 1. The Vegetational History of the Oaxaca Valley, by C. Earle Smith, Jr. Pages 39, 1 table, 2 maps, 10 plates. Part 2. Zapotec Plant Knowledge: Classification, Uses, and Communication About Plants in Mitla,

Oaxaca, Mexico, by Ellen Messer. Pages 149, 29 figures, 1 map, 10 plates. Prehistory and Human Ecology of the Valley of Oaxaca, Vol. 5. 1978. Price $8.
11. An Archaeological Survey of the Keban Reservoir Area of East-Central Turkey, by Robert E. Whallon. 1979. Pages 309, 211 figures, 20 tables, 2 plates. Price $10.
12. Excavations at Santo Domingo Tomaltepec: Evolution of a Formative Community in the Valley of Oaxaca, Mexico, by Michael E. Whalen. Prehistory and Human Ecology of the Valley of Oaxaca, Vol. 6. 1981. Pages 225, 38 tables, 58 figures, 73 plates. Price $13.
13. An Early Town on the Deh Luran Plain: Excavations at Tepe Farukhabad, edited by Henry T. Wright. 1981. Pages 462, 99 figures, 96 tables, 21 plates. Price $15.
14. Prehispanic Settlement Patterns in the Southern Valley of Mexico: The Chalco-Xochimilco Region, by Jeffrey R. Parsons, Elizabeth Brumfiel, Mary H. Parsons, and David J. Wilson. 1982. Pages 504, 128 figures, 115 tables, 40 maps, 31 plates. Price $16.
15. Monte Albán's Hinterland, Part 1: The Prehispanic Settlement Patterns of the Central and Southern Parts of the Valley of Oaxaca, Mexico, by Richard E. Blanton, Stephen Kowalewski, Gary Feinman, and Jill Appel. Prehistory and Human Ecology of the Valley of Oaxaca, Vol. 7. 1982. Pages 506, 43 tables, 139 figures. Price $20.
16. A Fuego y Sangre: Early Zapotec Imperialism in the Cuicatlán Cañada, Oaxaca, by Elsa M. Redmond. Studies in Latin American Ethnohistory & Archaeology, Vol 1. 1983. Pages 216, 75 figures, 46 tables, 42 plates. Price $15.
17. Irrigation and the Cuicatec Ecosystem: A Study of Agriculture and Civilization in North Central Oaxaca, by Joseph W. Hopkins, III. Studies in Latin American Ethnohistory & Archaeology, Vol. 2. 1984. Pages 148, 13 figures, 3 tables, 16 plates. Price $15.
18. Aztec City-States, by Mary G. Hodge. Studies in Latin American Ethnohistory & Archaeology, Vol. 3. 1984. Pages 166, 64 figures, 30 tables. Price $15.
19. Early Neolithic Settlement and Society at Olszanica, by Sarunas Milisauskas. 1986. Pages 319, 160 figures, 153 tables, 51 plates. Price $20.
20. Chipped Stone Tools in Formative Oaxaca, Mexico: Their Procurement, Production and Use, by William J. Parry. Prehistory and Human Ecology of the Valley of Oaxaca, Vol. 8. 1987. Pages 178, 42 tables, 52 figures, 20 plates. Price $18.
21. Conflicts Over Coca Fields in XVIth-Century Perú, by María Rostworowski de Diez Canseco. Studies in Latin American Ethnohistory & Archaeology, Vol. 4. 1988. Pages 314, 21 figures, 2 tables. Price $19.50.
22. Agricultural Intensification and Prehistoric Health in the Valley of Oaxaca, Mexico, by Denise C. Hodges. Prehistory and Human Ecology of the Valley of Oaxaca, Vol. 9. 1989. Pages 146, 6 figures, 42 tables, 4 appendices. Price $16.
23. Monte Albán's Hinterland, Part II: Prehispanic Settlement Patterns in Tlacolula, Etla and Ocotlán, the Valley of Oaxaca, Mexico, by Stephen Kowalewski, Gary Feinman, Laura Finsten, Richard Blanton and Linda Nicholas. 1989. Pages 1146 (in 2 vols.), 141 figures, 116 tables, 8 plates, 9 appendices. Price $40.

Technical Reports

2. LONGTERM and PEAKSCAN: Neutron Activation Analysis Computer Programs, by Thomas Meyers and Mark Denies. Contributions in Computer Applications to Archaeology, No. 2. 1972. Pages 76, 43 pages computer output, 2 figures. Price $1.
3. Data on the Abnormal Hemoglobins and Glucose-6-Phosphate Dehydrogenase Deficiency in Human Populations, by Frank B. Livingstone. Contributions in Human Biology, No. 1. 1973. Pages 289. Price $2.50.
4. An Archaeological Investigation on the Loboi Plain, Baringo District, Kenya, by William R. Farrand, Richard W. Redding, Milford H. Wolpoff, and Henry T. Wright. Research Reports in Archaeology, No. 1. 1976. Pages 59, 10 figures. Price $3.50.
5. Digging for Gold: Papers on Archaeology for Profit, edited by William K. Macdonald. Research Reports in Archaeology, No. 2. 1976. Pages 86. Price $3.50.
6. An Investigation of Ethnographic and Archaeological Specimens of Mescalbeans (*Sophora seconiflora*) in American Museums, by William L. Merrill. Research Reports in Ethnobotany, No. 1. 1977. Pages 167, 3 figures, 3 tables, 25 plates. Price $5.
7. Excavations at Quachilco: A Report on the 1977 Season of the Palo Blanco Project, by Robert D. Drennan. Research Reports in Archaeology, No. 3. Pages 81, 18 figures. Price $4.
10. Archaeological Investigations in Northeastern Xuzestan, 1976, edited by Henry T. Wright. Research Reports in Archaeology, No. 5. 1979. Pages 140, 52 figures, 20 tables. Price $6.

11. Prehistoric Social, Political, and Economic Development in the Area of the Tehuacan Valley: Some Results of the Palo Blanco Project, edited by Robert D. Drennan. Research Reports in Archaeology, No. 6. 1979. Pages 260, 46 figures, 26 tables. Price $6.50.
12. Late Prehistoric Bison Procurement in Southeastern New Mexico: The 1978 Season at the Garnsey Site, by John D. Speth and William J. Parry. Research Reports in Archaeology, No. 7. 1980. Pages 384, 39 figures, 32 tables, 34 plates. Price $9.
14. Archaeological Settlement Pattern Data from the Chalco, Xochimilco, Ixtapalapa, Texcoco and Zumpango Regions, Mexico, by Jeffrey R. Parsons, Keith W. Kintigh, and Susan Gregg. Research Reports in Archaeology, No. 9. Pages 222. Price $8.
15. The Garnsey Spring Campsite: Late Prehistoric Occupation in Southeastern New Mexico, by William J. Parry and John D. Speth. Research Reports in Archaeology, No. 10. Pages 228, 24 figures, 27 tables, 24 photos. Price $8.
16. Regional Archaeology in the Valle de la Plata, Colombia: A Preliminary Report on the 1984 Season of the Proyecto Arqueológico Valle de la Plata, edited by Robert D. Drennan. Research Reports in Archaeology, No. 11. 1985. Pages 195 (including complete Spanish translation), 43 figures, 16 tables. Price $8.
17. Zooarchaeology of Six Prehistoric Sites in the Sierra Blanca Region, New Mexico, by Jonathan C. Driver. Research Reports in Archaeology, No. 12. 1985. Pages 103, 29 tables, 8 figures, 1 appendix. Price $5.
18. The Henderson Site Burials: Glimpses of a Late Prehistoric Population in the Pecos Valley, by Thomas R. Rocek and John D. Speth. Research Reports in Archaeology, No. 13. 1986. Pages 348, 118 figures, 63 tables. Price $13.
19. Medicinal Plants of Native America, by Daniel E. Moerman. 1987. Pages 912 (in 2 vols.). Price $30.
20. Late Intermediate Occupation at Cerro Azul, Peru, by Joyce Marcus. 1987. Pages 112, 70 figures. Price $8.
21. The Inscriptions of Calakmul: Royal Marriage at a Maya City in Campeche, Mexico, by Joyce Marcus. 1987. Pages 205, 65 figures, 7 tables. Price $8.

Special Publications

The Williams Collection of Far Eastern Ceramics, Chinese, Siamese, and Annamese Ceramic Wares Selected from the Collection of Justice and Mrs. G. Mennen Williams in the University of Michigan Museum of Anthropology, by Kamer Aga-Oglu. 1972. Pages 73, 85 black and white photographs. Price $4.

The Williams Collection of Far Eastern Ceramics—Tonnancour Section, by Kamer Aga-Oglu. 1975. Pages 185, 183 black and white photographs, 18 color photographs. Price $8.